T0296794

Troubled Waters

Ocean Science and Governance

More than seventy per cent of the planet's surface is covered by the ocean and an estimated 40% of the world's human population lives in the coastal zone. The oceans are an international commons forming an essential part of our environment: the air we breathe, the water we drink, the weather and the climate. In addition, we rely on the oceans for food, to carry 80% of our trade, to absorb our wastes, as a source of energy, and as part of our culture and enjoyment.

This volume has been compiled to commemorate the 50th anniversary of the Intergovernmental Oceanographic Commission of UNESCO, which for half a century has been the UN organization responsible for fostering intergovernmental cooperation on global ocean science. It draws on the experience of 30 international experts to look at how governments use science to establish ocean policies, with chapters ranging from the history of ocean management to current advances in marine science, observation and management applications, and the international agencies that coordinate this work.

With a focus on key topical issues such as marine pollution, exploitation and hazards, *Troubled Waters* reflects on past successes and failures in ocean management and emphasizes the need for knowledge and effective government action to direct decisions that will ensure a sustainable future for this precious resource. Illustrated with dramatic, full-colour images, it is essential reading for researchers, students, policy makers and managers of the marine environment, and also provides an attractive and accessible overview for anyone concerned about the future stewardship of our oceans.

Geoff Holland was awarded an MSc in hydrodynamics and aerodynamics from London University, and served as a Scientific Officer for the UK Government Hydraulic Research Station in Wallingford (1957–67). He emigrated to Canada and spent 32 years in ocean science with the Canadian Government, culminating with a position as Director General, Ocean Sciences and Services. During this time he served on many national committees dealing with issues such as climate, energy, offshore resources, pollution, ocean data buoys and remote sensing. Amongst his intergovernmental duties were: Chairman of the London Convention (1972) of the International Maritime Organization, 1985–1989, overseeing the discussions leading to the ban on the ocean disposal of low-level radioactive waste; Chairman of the Intergovernmental Oceanographic Commission (IOC), 1995–1999; and Chairman of the Arctic Ocean Sciences Board, 1994–1997. Mr. Holland retired in 1999, but has remained active, including an appointment as a Canadian 'Ocean Ambassador' by the Minister for Fisheries and Oceans.

David Pugh was awarded a PhD in geodesy and geophysics from the University of Cambridge in 1968 before joining the Proudman Oceanographic Laboratory in Merseyside. In 1984 he became Head of Oceanography, Hydrology and Meteorology, Science Division, for the UK National Environment Research Council. He has also served as Secretary to the United Kingdom Government Committee on Marine Science and Technology and was elected President of the IOC (2003–2007), having previously been the Founding Chairman of the IOC Global Sea Level network, GLOSS. Dr Pugh maintains an active programme of research associated with Liverpool University and the Proudman Oceanographic Laboratory. His interests include tides, surges, mean sea level, coastal management and climate change, the economics of marine activities related to GDP, and the history of sea level, and he is the author of two other books.

The Intergovernmental Oceanographic Commission (IOC) of UNESCO celebrates its 50th anniversary in 2010. Since taking the lead in coordinating the International Indian Ocean Expedition in 1960, the IOC has worked to promote marine research, protection of the ocean, and international cooperation. Today the Commission is also developing marine services and capacity building, and is instrumental in monitoring the ocean through the Global Ocean Observing System (GOOS) and developing marine-hazards warning systems in vulnerable regions. Recognized as the UN focal point and mechanism for global cooperation in the study of the ocean, a key climate driver, IOC is a key player in the study of climate change. Through promoting international cooperation, the IOC assists Member States in their decisions towards improved management, sustainable development, and protection of the marine environment.

Dr Wendy Watson-Wright
Executive Secretary
Intergovernmental Oceanographic Commission of UNESCO

United Nations
Educational, Scientific and
Cultural Organization

Organisation
des Nations Unies
pour l'éducation,
la science et la culture

TROUBLED WATERS

Ocean Science and Governance

Edited by

GEOFF HOLLAND

and

DAVID PUGH

This publication has benefitted from substantial financial support made available by the Belgian and Canadian Governments

Shaftesbury Road, Cambridge CB2 8EA, United Kingdom

One Liberty Plaza, 20th Floor, New York, NY 10006, USA

477 Williamstown Road, Port Melbourne, VIC 3207, Australia

314–321, 3rd Floor, Plot 3, Splendor Forum, Jasola District Centre, New Delhi – 110025, India

103 Penang Road, #05–06/07, Visioncrest Commercial, Singapore 238467

Cambridge University Press is part of Cambridge University Press & Assessment, a department of the University of Cambridge.

We share the University's mission to contribute to society through the pursuit of education, learning and research at the highest international levels of excellence.

www.cambridge.org
Information on this title: www.cambridge.org/9780521765817

First published 2010

A catalogue record for this publication is available from the British Library

Library of Congress Cataloging-in-Publication data
Holland, Geoff, 1935–
Troubled waters : ocean science and governance / Geoff Holland, David Pugh.
p. cm.
ISBN 978-0-521-76581-7 (hardback)
1. Marine sciences – Government policy. 2. Marine ecology – Government policy. 3. Marine pollution – Prevention – International cooperation. I. Pugh, D. T. II. Title.
GC28.H65 2010
551.46–dc22 2010015148

ISBN 978-0-521-76581-7 Hardback

Contents

Advisory Board

Foreword

Which part of our planet is the most shared, used, exploited yet the least well known, protected … and surveyed?

Without doubt, our Oceans! They host all our causes for concern and demonstrate the urgent need for concrete, effective action.

All the far-reaching changes that we have experienced in recent years have been typified by the fact that they have taken place over a time scale which, up until now, nature had only experienced during terrestrial or planetary catastrophes, at a time when Man and his civilization had not yet seen the light of day.

Man has harnessed emerging advanced technologies to master nature – believing that he is not bound by original constraints – and now those same technologies have triggered mechanisms which could threaten the very existence of us 'intelligent animals'. The history of humanity is in the process of being rewritten right now. Managing this change requires our technical civilization to coexist in harmony with the ecological system. This shift requires an in-depth understanding of our planet.

The mechanics of the oceans are proving to be incredibly complex and require us to act globally in terms of both geography and theme. They require all the exact, natural and human sciences to be harnessed in the name of conservation.

Applying the principle of precaution, before it was a buzzword, our ancestors sensed an overwhelming need for international collaboration founded on empirical, then scientific knowledge. In 1908, under the auspices of the International Geography Congress, my great-great grandfather Prince Albert I founded a special Mediterranean commission which subsequently became the CIESM, Mediterranean Science Commission, thereby making his ardent ambition for effective intergovernmental cooperation come true.

In 1899, when the international community expressed a desire to have a coherent cartographic tool, he decided to commission the General Bathymetric Chart of the Oceans (GEBCO) and in 1921 he hosted the International Hydrographic Office in Monaco. GEBCO and CIESM are channels which, for half a century, have enabled my Country to work in conjunction with the Intergovernmental Oceanographic Commission.

For my part, without neglecting the many other sensitive areas of our planet, both terrestrial and marine, it is the polar regions that caught my attention. They suffer from the combined effects of local environmental changes and the causes of global upheaval. It is astonishing to say the least that these regions, which, for several years, have been affected by human activities, even those far away – by which I mean in particular radioactive or chemical pollutants – will in turn have a catastrophic impact on the lifestyles of far away populations, in particular those located in coastal or island regions. These changes are precisely fuelled by ocean and atmospheric forces.

One hundred years ago Prince Albert I had already stressed the union of these powers on the facade of the Oceanographic Museum in the form of an allegory entitled 'Truth unveiling the forces of the World to Science'.

It is in this context – where ecological phenomena outlast the political lifetime of decision-makers, where the frontiers of nations are no longer barriers - that the principles of harmonization and intergovernmental collaboration take on their full value.

Over the past 50 years the International Oceanographic Commission (IOC) has affirmed this vocation. I would like to pay homage to its works, past and present. They are all the more worthy of merit as it is difficult to convince managers of the importance, for the survival of our civilizations, of the health of natural phenomena in the depths of oceans, far from our contemporaries' eyes.

This strength of conviction has borne its fruits and I am delighted to observe the gradual consensus which has grown up between the various sections of our societies.

Although it is relatively easy to see what is happening on the surface of the oceans, their abysses are sadly subject to harmful effects. Quite recently, a conference of experts invited to the Principality under the auspices of the IOC demonstrated how the deepest layers of life in our seas, the most intimate biological processes, are disrupted by imbalances of CO_2, a gas to which life owes its conquest of the terrestrial environment. The acidification of our oceans literally burns the very organisms partially designed to store it.

This is a dramatic demonstration of the way in which an ecological system is packaged, insidious chaos insinuating itself into the very heart of the regulation mechanisms, into life itself. It is also an urgent call to take action to ensure that we do not disappoint our marine ancestors or forget our sea!

HSH Prince Albert of Monaco

Palais de Monaco, 3rd of November 2009

Acronyms

The following acronyms are those found most often throughout the book. Several acronyms, used and defined in individual chapters, are not included here in the interests of brevity.

CBD	Convention on Biological Diversity
CCCO	Committee on Climatic Changes and the Ocean
CLCS	Commission on Limits of the Continental Shelf
CLIVAR	Climate Variability and Predictability Study
CSD	Commission on Sustainable Development
CZM	Coastal Zone Management
DOALOS	Division for Ocean Affairs and the Law of the Sea (UN)
EEZ	Exclusive Economic Zone
FAO	Food and Agriculture Organization
FCCC	Framework Convention on Climate Change
GCOS	Global Climate Observing System
GEBCO	General Bathymetric Chart of the Oceans
GEF	Global Environment Facility
GEOSS	Global Earth Observation System of Systems
GESAMP	Joint Group of Experts on the Scientific Aspects of Marine Environment Protection
GIPME	Global Investigation of Pollution in the Marine Environment
GLOSS	Global Sea Level Observing System (IOC)
GOOS	Global Ocean Observing System
IAEA	International Atomic Energy Agency
ICAM	Integrated Coastal Area Management
ICES	International Council for the Exploration of the Sea
ICP	United Nations Open-ended **Informal Consultative Process** on Oceans and the Law of the Sea
ICSPRO	Inter-Secretariat Committee on Scientific Programmes Relating to Oceanography
ICSU	International Council for Science
IDOE	International Decade of Ocean Exploration
IGBP	International Geosphere Biosphere Programme
IGOSS	Integrated Global Ocean Station System
IIOE	International Indian Ocean Expedition
IHO	International Hydrographic Organization
IMO	International Maritime Organization

IOC	Intergovernmental Oceanographic Commission (UNESCO)
IOCARIBE	IOC Sub-Commission for the Caribbean and Adjacent Regions
IODE	International Oceanographic Data and Information Exchange (IOC)
IPCC	Intergovernmental Panel on Climate Change (UNEP-WMO)
JCOMM	Joint WMO-IOC Technical Commission for Oceanography and Marine Meteorology
JGOFS	Joint Global Ocean Flux Study
NEPAD	New Partnership for Africa's Development
OBIS	Ocean Biogeographic Information System
OSPAR	Oslo and Paris Convention
PEMSEA	Partnerships in Environmental Management for the Seas of East Asia
PICES	North Pacific Marine Science Organization
SCAR	Scientific Committee on Antarctic Research
SCOR	Scientific Committee on Oceanic Research (ICSU)
TOGA	Tropical Ocean and Global Atmosphere (programme) (WCRP)
UN	United Nations
UNCED, UN	Conference on Environment and Development, Rio de Janeiro 1992
UNCLOS	United Nations Convention on the Law of the Sea
UNDP	United Nations Development Programme
UNEP	United Nations Environment Programme
UNESCO	United Nations Educational, Scientific and Cultural Organization
WCRP	World Climate Research Programme (WMO/ICSU/IOC)
WESTPAC	IOC Sub-Commission for the Western Pacific
WMO	World Meteorological Organization
WOCE	World Ocean Circulation Experiment (WCRP)
WSSD	World Summit for Sustainable Development, Johannesburg 2002

Part I Introduction

It is now commonly recognized that the human race is capable of changing the environment of planet earth. There are many different views about how little or how much; whether changes can be good or bad; or even whether we should be complacent or concerned. The continuance of this debate implies a fundamental lack of sure knowledge about the relationship between our civilization and the world we live in. Our future existence depends on our ability to understand and to manage future planetary changes. Our habitable space on earth is sandwiched between a barren vacuum above and an impenetrable rock barrier below, akin to the skin of water on a wet balloon. In the face of this vulnerability, it is a sobering thought that our species is so ignorant of the consequences of its actions.

The oceans cover nearly three-quarters of the planet's surface. They form an essential part of our environment, the air we breathe, the water we drink, the weather and the climate. In addition, we rely on the oceans for food, to carry 80% of our trade, to absorb our wastes and to be part of our culture and enjoyment. Since the ratification of the Law of the Sea in 1994, jurisdiction over large tracts of the coastal waters, stretching out to two hundred nautical miles and more, have been accorded to the Coastal States. However, much of the ocean waters remain an international commons and the ocean waters and their inhabitants recognize no man-made boundaries, flowing at will from one jurisdiction to another.

For all their importance to our society and the life on our planet, the oceans are still relatively unexplored. Until the latter half of the last century, our ocean science observations were mostly restricted to what could be gleaned from instruments and nets lowered over the side of research vessels. Today, modern technology in acoustics, underwater vehicles, automated floats and satellites are overcoming our lack of information, but there remains so much more to be done. Information is derived from observation and the possession of knowledge to interpret what is recorded. It is a continuum that is constantly changing.

Humankind has enjoyed many benefits from this changing ocean. Low-cost transport, an endless supply of food and somewhere to dump unwanted waste have been benefits from time immemorial. Various national economic assessments of the contributions of marine-related activities to national economies have typically shown that 5% of GDP, or in some

Figure 0.1 A blue sea and a calm voyage is the hope for every mariner.
© Carina-Foto/Shutterstock.com.

cases even more, is directly due to marine activities: increasingly these involve leisure industries, coastal and cruising; and of course world trade is dominated by marine transport. Indirectly the economic impacts of the sea for us are much greater than this. Even a decade ago few people recognized the important ecosystem services the oceans provide for maintaining a healthy planet. Now it is widely recognized that these services include carbon sequestration and climate regulation, primary production, and cultural, intellectual and spiritual stimulation. For economists the new challenge is to express the value of these services in traditional monetary terms. For governments, the challenge is to sustain these services for a viable future.

The need for ocean scientists and ocean policy makers to work together is clear. This book has that interaction as its central theme. Several authors show that there are close links between government policies and scientific possibilities. Scientists can explain the range of possible futures. Governments help to make selected futures happen. In marine science we may advise on ocean management possibilities, and

Figure 0.2 Storms at sea can be forecast and prepared for. The future of the oceans depends on planning now for a sustainable future. © Olly/Shutterstock.com.

governments may choose to take that advice. Or they may not, choosing in the face of uncertainty and societal pressures to do something different or nothing at all. This symbiosis between governments and science is often unrecognized on both sides, without closer examination of the mutual expectations, which are often very different.

Good ocean governance and wise stewardship require accurate and timely information. Politicians and governments want their environmental decisions to be based on 'good science'. Translated, that means that there should be scientific consistency and consensus, and an assumption that the actions recommended will be effective. Cynically, if there are unseen consequences of their decisions and actions, governments and politicians have someone else to blame. Unfortunately for them, science seldom delivers absolute certainty. Governments want marine legislation and regulation to be effective. They want operational systems, for example storm surge warnings and tsunami warning systems, to protect their citizens, and at reasonable cost. And inevitably they want wealth-creating businesses to continue to the maximum extent possible, to ensure jobs, individual prosperity and national economic success.

Scientists have a different agenda. Scientists need funds to carry out their research and ocean research is inherently expensive. Therefore it is governments that fund most marine science either directly, or indirectly. Good science requires conclusions to be independent of the expectations of the funding sources, and scientists need freedom to apply their professional skills without interference. Governments naturally desire control of expenditures and seek to ensure that research results lead to more efficient and effective services for their constituents. While governments need operating systems that can feed information into established management regimes, scientists are concerned that the results of labours will lead to an erosion of their scarce research funds to pay for the on-going expenses of such systems.

Scientists expect that their conclusions and recommendations will be taken seriously, and not dismissed because they are politically difficult. And above all, scientists want politicians to accept and acknowledge publicly that scientific assessments are inevitably expressed in terms of probabilities. If a hypothesis is not proven at the 95% level, in no way does that signify that the reverse hypothesis has been proved. Not every politician and few members of the general public appreciate that.

Another point to be made is that the term 'government' cannot be taken to imply a shared equality in capacity or capability. The lack of needed local scientific skills in many parts of the world calls for scientific cooperation, and for the transfer of expertise and technology. Many nations that now possess jurisdiction over the ocean bordering their coast have neither the ability nor the resources to manage their newly accorded territory. Without assistance, these disadvantaged nations cannot expect to protect the marine environment or profit from its bounty. The Law of the Sea obligates the ratifying nations to contribute to the transfer of technology, but those articles have so far been largely neglected, emphasizing once again the need for intergovernmental action.

When considering the roles of governments and government-funded science, the role of non-governmental organizations in coordinating ocean research and in raising public awareness must not be overlooked. The Internet has created new ways for groups to lobby and influence popular opinions. Global pressure groups can lobby internationally in a very effective way that can bypass the nation state and its decision-making processes. Governments may individually deprecate this, but collectively they may also realize the value and status of these independent voices, helping to translate the results into policies, decisions and actions.

This book has been prepared on the occasion of the fiftieth anniversary of the Intergovernmental Oceanographic Commission of UNESCO. The IOC is the United Nations organization dedicated to ocean science and services. The Member States, as represented at the IOC, invited us to lead in the preparation of a book that would explore some aspects of the relationships between science and governments. We have done this by approaching the authors of the individual chapters in a personal capacity, under the guidance of our Advisory Board.

Figure 0.3 The wealth of the oceans includes food as well as more exotic and beautiful things. © Anyka/Shutterstock.com.

Authors were asked to consider the issues within a global context. Although many of the achievements and weaknesses of the IOC are reflected in the following chapters, the IOC is a relatively small part of the world interest in the oceans. It is, however, a forum at which Member States have collectively expressed their concerns for the future of the oceans. They know that the oceans are a key part of the global environment, dealing with the health, climate, environment, economy and culture of our society. This book is one way of transmitting their message to the World. Certainly the role of the IOC is important and quite naturally references to its work will be prominent throughout these pages, but the story and the messages are necessarily much broader, and are addressed here to a wider audience.

It is no longer an indulgent dream that governments may come together and act in a coordinated way to understand the ocean and agree on its stewardship. It is now a necessity. For generations we have been lulled into a sense of security by the vastness of the oceans and their seemingly impregnability to our actions, but that sense of security is now seen to be an illusion.

It is true that governments have long accepted some responsibility for the oceans and have worked to apply science to the understanding and, they expect, to the eventual management of the oceans. However, their efforts have lacked urgency, and have been relatively small compared to the size of problems to be tackled. In many cases, national responses have been partial and inadequate. Neither the ocean environment, nor the creatures that inhabit it, recognize national boundaries and ocean issues need to be tackled on scales that transcend national, regional and ocean basin scales. Even very local issues can be ubiquitous in nature and can profit from observations and research in other areas.

This is not a book that covers the issues of oceans, science and governments in a comprehensive way; in fact it would be pointless to try to cover all ocean issues in a

Figure 0.4 The path to the sea is inviting, but the challenges are immense. © Thorsten Rust/ Shutterstock.com.

single volume. Still less does it offer immediate and easy solutions to the problems that lie ahead. It is a book in which individual practitioners with long experience give their impressions on the progress of science over the past 50 years, towards the understanding of ocean issues and how scientific efforts are being translated into information for management and decision-making. Many of the chapters reflect the IOC connection using examples drawn from that source, but it is the overall intention to look critically and widely at many of the aspects of oceans and intergovernmental cooperation over the past half-century and earlier, and where possible to record the lessons learned. The overall message is that governments need to work together with much greater urgency, to address the many natural and man-made issues concerning the ocean; they need to understand better the role that ocean science can play and they need to develop much stronger ocean governance mechanisms to profit from the knowledge obtained.

As editors we wish to thank all those who contributed: our formal and informal advisors, the several authors who produced excellent copy without excessive editorial bullying, and the several suppliers of colourful images which make the book a more pleasant read as well as frequently illustrating important points. These sources are acknowledged as appropriate in the text. We thank the UK Government for help with the images. Inevitably there will be some duplication, but this reflects the importance of the issues as seen from different perspectives, and we have not attempted to influence this.

Finally, we thank the Member States of the IOC for inviting us to produce this imperfect, incomplete, but we hope stimulating volume. It has been at times a demanding challenge, but ultimately, a satisfying and rewarding and worthwhile undertaking. We hope that our readers will agree.

Part II The Global Context: Preface

Over the past half-century science has shown that even the most remote parts of the ocean are in some way altered by human activity. This realization has been gradually spreading among the general public in all countries, as has the knowledge that the oceans supply to the global ecosystem a series of environmental services that make life possible on this planet. We are part of that ecosystem, and if there is to be a viable future, then collectively we must manage and control our activities so that the oceans are healthy, and their role in maintaining a healthy planet is sustainable.

Unfortunately, public awareness, no matter how widespread, is not alone a guarantee of improvement, or even of stopping the progressive deterioration of the oceans that cover some 70% of the earth's surface. That demands both political will and effective intergovernmental mechanisms to make decisions for the common good, and to ensure that these decisions are implemented, if necessary by common policing arrangement.

The intergovernmental mechanisms for ocean management are weak and little changed over many decades; as such they are not adequate for the responsibilities they carry. And it is a tragedy of the global commons that individual governments, in the absence of overwhelming evidence and social pressures, will always be reluctant to make difficult decisions, however necessary they may be for the health of the global ocean. Fish do not have a vote. The United Nations Convention on the Law of the Sea, which came into force in 1992, is still not ratified by some major governments, nor is there any effective mechanism for monitoring, much less policing, its implementation. While governments have occasionally shown a willingness to use military force to protect their designated seas, the High Seas have no such protection.

When the United Nations was being established over 60 years ago, along with a series of specialized agencies, ocean interests were weak and scattered among many organizations. For many of these the oceans were, and remain, a non-central responsibility in their overall remit. More recently several voices have called for a new UN agency for the oceans. But creation of such an agency would be politically impossible today in the face of pressures from many governments to reduce not extend the UN system, and of course from the existing agencies, who would not want to lose their interest, however partially pursued, in the global ocean.

In this first section of the book we have two chapters dealing with the multiplicity of mechanisms for interaction among national governments, mainly within the UN system, and some measures that have been taken in the past decade to begin to address the need for a coordinated approach among governments. To be fair, the problems of ocean governance are recognized in many quarters, but the possible actions that can be taken are woefully slow and inadequate, compared with the scale and urgency of the problems and the scale of the remedial measures needed. Even the systematic assessment of the state of the global oceans, called for in 2002 at the World Summit on Sustainable Development, has been delayed over the last 8 years, in part due to the objection of a few nations based on perceived national interests.

Science has a role to play in both informing public opinion about the need for action, and in advising governments, individually and collectively, on the measures that are necessary, and which will be effective. Within the United Nations Convention on the Law of the Sea (UNCLOS) there are provisions for encouraging marine scientific research, as a common good; there are also regulations that control the pursuit of marine science in the Exclusive Economic Zones (EEZs) of coastal states, the implications of which are still being negotiated. The chapter on UNCLOS looks at some of the developments in the application of UNCLOS for marine science.

Over the past half-century scientists have planned and implemented some major ocean-wide, and later global, experiments. The Intergovernmental Oceanographic Commission, established as a body having functional autonomy within UNESCO, in 1960, has been the pioneering agency within the UN system for intergovernmental marine scientific cooperation. Various aspects of the IOC activities, achievements and disappointments will be covered later in this book. This section includes an account of the early days and evolution of the IOC: the vision and some of the perhaps inevitable difficulties. These early and faltering steps are the footprints that will eventually make a road to a fuller scientific understanding of the oceans and their importance for all humankind.

Let there be no mistake: the oceans and their future are not just of interest to a few more prosperous nations. Many developing nations have extensive coastlines, and coastal populations whose livelihood and security from the sea will be increasingly threatened. And it is the poorest countries and people who will suffer first and most acutely from the negative effects of climate change, a change that is at present moderated by ocean influences. These countries sometimes lack the basic national capabilities to commission, access, resource and implement remedial measures to sustain their coastal waters in a practical way. Understanding and protecting the global ocean needs the participation of all coastal nations if the monitoring of oceans and ocean changes is to be effective for the benefit of all humankind. Here, a chapter outlines some lessons learned, over 50 years, in how to transfer scientific and technical marine skills both effectively and sustainably.

1 For the ocean

PATRICIO A. BERNAL

Patricio Bernal was Executive Secretary of the Intergovernmental Oceanographic Commission, and Assistant Director-General of UNESCO from 1998 to 2009. He studied at the University of Chile and Scripps Institution of Oceanography. From 1990 to 1994, he was the Executive Director of the Fisheries Development Institution of Chile, then Under-Secretary of State for Fisheries in the Ministry of Economics. His scientific interests include marine ecosystem interactions and the impacts of global climate change.

The ultimate global commons

The Ocean is the ultimate global commons. Life originated on earth because it has the ocean: water in liquid state on its surface. Although curiosity has pushed humankind to search for life in outer space, there is still no other planet with life in the known universe.

Life on earth originated in the margins of the primordial ocean and for millions of years evolved in this aquatic milieu. The ocean is a thin layer of fluid that plays an essential role in making the planet liveable: on average the radius of the planet is 6371 km and the ocean is 3733 m deep, i.e. a thickness of 0.058% or 6 ten thousandths of the radius. The ocean is to the earth thinner than the skin is to an apple. The ocean is the ultimate global commons because it provides essential ecological services to all humankind, making life possible on our planet. For example, marine plants produce 36 billion tons of oxygen annually, which is estimated to be equal to 70% of the oxygen in the atmosphere. I cannot think of a more fundamental reason to assert that every form of life on the earth has a stake on the health of the ocean. Humanity, by mastering the technological power to disrupt its equilibrium, is especially responsible for its health.

Figure 1.1 CAPT Paladej Chareopool (RTARF), RADM Dato'Jamil Bin Osman (MAF), RADM Tay Kian Seng (RSN) and Air First Marshall Bonggas Somoring Sailaen (TNI) at the Inaugural Malacca Strait Patrols Information Sharing Exercise. © IOC/UNESCO.

Once, when I was explaining that due to the nature of fluids, the ocean should be considered to be one, singular ocean, there was a strong reaction from part of the audience, many of them lawyers, who immediately challenged the unity of the ocean concept and argued in favour of maritime spaces subject to dominant national interest, reflecting the vision of the ocean as open to the competition and dominion of nations, that is, a territorial space. Half of the lawyers in this world are trained within the Roman legal tradition, the other half under consuetudinary or traditional law. The commons is a concept that emerges naturally and is harmonious within consuetudinary law but does not fit easily in the Roman tradition where it becomes public good or public space.

Much more than fish and ships

There are alarming signs that the management systems that we have are insufficient to guarantee the integrity of the several natural systems that provide basic ecological services to humanity and the sustainability of living marine resources. Many of the uses that man makes of the ocean are having secondary effects that adversely impact the stability of natural processes in the ocean. Destruction of critical habitats along the coast is alarming, as human populations encroach onto the coastal zone. Destruction of deep ocean habitats is significant due to the secondary effect of fish trawling. Destruction of corals due to bad practices in the collection of fish for aquaria is still going on in several regions of the world. Massive accumulations of plastic in the central gyres of the Pacific Ocean are only recently being detected and studied. There is increasing frequency and abundance of dead-zones due to the exhaustion of oxygen by the arrival of vast quantities of chemicals used by, or originating in, industry, agriculture and animal husbandry and transported by rivers into the ocean.

Figure 1.2 Elisabeth Mann-Borgese was a Canadian expert on maritime law and policy, and an advocate for international cooperation. Image courtesy of IISD/Earth Negotiations Bulletin.

Through photosynthesis by microscopic plants in the surface layer of the ocean carbon dioxide is drawn from the atmosphere and oxygen is released. The exchange of oxygen and carbon dioxide (and other gases) has a profound effect on the earth's climate. Absorbing millions of tons of CO_2 every year – roughly one-third of total annual emissions – the ocean has already spared us from catastrophic climate change. But in doing so, its own intrinsic balances are being altered: it is becoming more acidic and has taken the largest fraction of the additional heat generated by anthropogenic greenhouse gases, something that might eventually alter the normal patterns of ocean circulation that are so essential for keeping the absorbed CO_2 from reuniting with the atmosphere for long periods, buying us time for finding the solutions.

We have an incomplete and piecemeal picture of what is happening to the ocean and there is an urgent need to change this and adopt corrective policies at the highest level possible. Too much is at stake to follow the path of least resistance. Powerful political leadership is needed.

A fluid spatial domain, a unique international space

The ocean is a unique international space that for many centuries remained almost fully outside any national jurisdiction and still today its largest fraction (64%) remains an international commons. The development of the ocean legal regime through the centuries is abundant in written theory and legal innovations, reflecting the evolving balance between the interests of nations, the availability of technology and the power to impose and maintain a status quo. In the following paragraphs I have drawn from Susan J. Buck's historical

Figure 1.3 Patricio Bernal discusses the future of the oceans with President Grímsson of Iceland. © IOC/UNESCO.

revision in her 1998 book, *The Global Commons, an introduction*. As early as the second century, reflecting their own position in the Mediterranean, the Romans had declared that the seas were *communes omnium naturali jure*, or common to all humankind. The growth in power and commerce by the mercantile-cities (by 1269 Venice was charging tolls from all vessels in the Adriatic Sea) and coastal nations made the control of coastal waters strategic and states started to lay claim on the oceans to protect their interests and project their influence. The most ambitious of these claims was that of Spain and Portugal, which in 1494 agreed in the Treaty of Tordesillas, under the authority of the Pope, to divide the whole world into two equal hemispheres: one under Portuguese and the other under Spanish dominion. In fact the untenable practical situation created by these treaties, in view of the growing influence of Dutch, English and French maritime presence, fuelled a vigorous discussion around the concepts of closed seas (*mare clausum*) and freedom of the seas (*mare liberum*) marking the origins of modern law of the sea. In 1602 the Dutch East India Company seized a Portuguese galleon in the Strait of Malacca, until very recently still one of the hotspots of world piracy, and the company commissioned Hugo Grotius to write a paper with the legal basis justifying that action. The famous piece 'Freedom of the seas' (*Mare liberum*) by Grotius is part of a larger opus 'On the law of the spoils' (*De jure praedae*), which in perfect symmetry was very quickly subject to rebuttal by an English jurist, John Selden, who at the request of King James I, wrote 'Mare Clausum: the right and dominion of the sea', in defence of the British seizure of Dutch cargoes off Greenland. The history of the law of the sea is long and fascinating. It is not my intention to summarize it here, only to emphasize that the ocean law regime is old, has evolved, has had to adapt to the power and political realities of the times, and will continue to change.

Figure 1.4 Stromatolites are the world's oldest fossils. © Monika Johansen/Shutterstock.com.

UNCLOS and the promises of its entry into force

Susan Buck coins the suggestive statement 'technology has caught up with desire' to describe the current situation where technology now exists for extracting value and for establishing and sustaining property rights from the vast spatial domains considered as global commons. The United Nations Convention on the Law of the Sea (UNCLOS) was adopted in December 1982 in Montego Bay, Jamaica, and entered into force on the 16th of November, 1994. According to Federico Mayor, then Director-General of UNESCO, this provided for the 'largest transfer of resources in the history of humanity' by defining the limit of the territorial sea as 12 miles, the adjacent zone to 24 miles, the continental shelf up to its natural limit and creating the Economic Exclusive Zones extending to 200 nautical miles offshore from straight-line baselines at the coast, with the possibility of extending it up to 350 miles under special circumstances. Excluding the deep-seabed mining regime contained in part XI and its subsequent Implementation Agreement, UNCLOS is the most recent and classic case of *enclosure* in international law; it encloses 36% of the world's ocean, including 90% of commercially exploitable fish and 87% of projected offshore oil reserves. The question remains, 'Has this transfer been effective for the protection of the ecological services and good management of the resources the ocean contains?'

UNCLOS provides an integrated legal framework on which to build sound and effective regulations to the different uses of the ocean. These have been implemented by the UN specialized agencies and programmes over the last 30 years. Nevertheless, severe limitations exist for monitoring and enforcing these regulations. National and international institutions for the ocean are fundamentally weak. They are usually compartmentalized on a sector by sector division of duties and responsibilities, leaving little room for integrated policy-making or addressing issues that cut across several domains.

The ocean environment is a fluid in constant flux. Because of the highly interconnected and dynamic nature of the ocean environment, human activities (or natural processes) – taking place locally at one site – can influence and affect the outcome of another activity (or natural process) at an adjacent or even a very distant location. Elizabeth Mann Borgese drew the consequences of these properties in terms of the key concepts of the classical legal order on land: 'the ocean forces us to think differently (...) many "terrestrial" concepts simply will not work in the ocean medium. These concepts include property in the Roman law sense, sovereignty in the sense of the "Westphalian era" and "territorial boundaries", which neither fish nor pollution will respect'. In her 1998 book *The Oceanic Circle, Governing the seas as a global resource*, a report written for the Club of Rome, Mann-Borgese explores a most ambitious and visionary picture on how to expand the new ocean law regime 'for the making of a genuinely new international/national political, legal and social order'.

What we face in the ocean is a problem of management. Sectors tend to define the problems they face as internal, and seek solutions exclusively from within the sector, when in fact many factors affecting the sector are being impacted by sector activities that lie outside it. Technically this compartmentalization externalizes the costs to other sectors and is globally inefficient. Somebody will have to pay the external cost either now or in future generations.

The sustainable use of the ocean and of its resources calls for the application of an integrated management regime. Despite the call contained in Chapter 17 of Agenda 21 (1992), which formulated a comprehensive prescription for the integrated development of the ocean environment, it is only recently, with the increased uses of the ocean and its resources, that the many shortcomings of the sector approach have become apparent.

Figure 1.5 In most Indonesian and Philippines coastal waters, fishermen use small boats like this to catch their fish. © Alessio Viora/Marine Photobank.

Although progress has been slow, there are important initiatives that have gained ground and kept the general process moving. Some of the recommendations made in 1998 by the Independent World Commission on the Ocean, chaired by Mario Soares the former president of Portugal, have been successfully implemented, such as the establishment of an Informal Consultative Process at the level of the General Assembly of the United Nations. This forum was conceived to provide an opportunity to discuss annually the Ocean agenda with the direct participation of governments, the specialized agencies and programmes of the UN and the representatives of NGOs and civil society.

Several countries have established and issued principles of National Ocean Policy, usually codified in public documents by a high-level authority, setting up standards for all the activities to be conducted in the ocean space under national jurisdiction, usually empowering cross-sectoral coordination. The IOC Technical Series 75 report, published in 2007, reproduces the National Ocean Policies from Australia, Brazil, Canada, China, Colombia, Japan, Norway, Portugal, Russian Federation and the USA. Many countries have developed plans and institutions aimed at progressing towards the integrated management of their coastal zones, starting with the application of spatial zoning schemes for the near-shore environment.

Successful cooperation exists at the regional level in several parts of the world, providing for the voluntary implementation of the principles of UNCLOS or Chapter 17 of Agenda 21, as well in the enforcement of international law within and beyond the limits of national jurisdiction. An example is the Partnerships in Environmental Management for the Seas of East Asia (PEMSEA), the very successful follow-up of a GEF Project started in 1994. The integrated management of the maritime spaces within the 200 mile EEZ is starting to be seen and is being addressed by some governments as a single challenge and opportunity, underpinning a significant ocean economy.

In China the ocean is estimated to be equivalent to 4–11% of the GNP, and therefore a key factor for their development. Several of the 17 conventions of Regional Seas have achieved a significant level of agreement on policies, have built effective institutions and regularly implement functional tasks, such as conducting regional assessments.

In the Mediterranean basin, the Barcelona Convention system created in 1996 a regional Commission for Sustainable Development with an innovative structure and function. This comprises 46 members: the 22 contracting parties and 24 rotating representatives of local authorities, socio-economic actors and non-governmental organizations working in the fields of environment and sustainable development. Moreover, the representative of the parties is not necessarily always the minister of the environment, as is traditional in the Regional Seas Conventions.

The United Nations

Similar to the national level, where almost all ministries of a government have some function or authority related to ocean affairs, in the United Nations system almost all specialized agencies and programmes are involved in ocean affairs. I am in debt to the late Elizabeth Mann-Borgese for guiding me in this review. The International Maritime Organization (IMO), the International Seabed Authority (ISA) and the Intergovernmental Oceanographic Commission (IOC of UNESCO) are exclusively devoted to ocean affairs: IMO for shipping, ISA for seabed mining and IOC for ocean sciences and ocean services.

The United Nations Educational Scientific and Cultural Organization (UNESCO), the Food and Agriculture Organization (FAO) and the United Nations Environment Programme (UNEP) have broader mandates, including divisions dealing with ocean affairs. The FAO has responsibility for fisheries and aquaculture and the UNEP for regional seas and marine environment. UNESCO having eliminated its Marine Sciences Division in 1990 in favour of concentrating ocean sciences under the IOC, still maintains other programmes focusing on small islands development states (SIDS), culture (the secretariat for the Underwater Cultural Heritage Convention) and Education (Division on Education on Sustainable Development).

Other UN organizations are also involved with the ocean, such as the World Meteorological Organization (WMO) dealing with ocean–atmosphere interaction, marine meteorology and climate and its implications, the International Atomic Energy Agency (IAEA) for nuclear marine pollution, the United Nations Industrial Development Organization (UNIDO) with industrial marine technology, the

Figure 1.6 Local fishermen in Anilao, Philippines, use a simple hook and line to catch squid, a sustainable fishing practice. © Peri Paleracio/Marine Photobank.

International Labour Organization (ILO) for the protection of maritime workers in the shipping and fisheries industries, the World Health Organization (WHO) for ocean-related health problems and food-safety, the United Nations Development Programme(UNDP) and the World Bank, financing the sustainable development of ocean and coasts.

Several Divisions of the central UN Secretariat also play a role: the Division of Economic and Social Affairs (UN-DESA) acts as the secretariat for the Commission on Sustainable Development, coordinating programmes for coastal management, small island development states and ocean. The Division of Ocean Affairs and the Law of the Sea (UN-DOALOS) acts as the secretariat for UNCLOS, the Commission on the limits of the continental shelf and by default for any other meeting on oceans that is organized under the central UN system in New York, as is the case today for the Informal Consultative Process on Oceans.

Although the streamlining of agencies and programmes and the harmonization of policies has long been in the agenda of the UN and was entrusted in the past to the Administrative Coordination Committee (ACC), comprising all heads of agencies and programmes and chaired by the UN Secretary-General, little progress has been made. Perhaps realizing the limits of the inter-secretariat level to tackle this type of institutional policy formulation, one of the reform measures of former Secretary-General Kofi Annan was to abolish the ACC itself, and therefore all its subsidiary bodies, including the Inter Agency Committee on Sustainable Development and its Sub-Committee on Oceans and Coastal Areas (IACSD-SOCA), created after the Rio de Janeiro Conference on Environment and Development. Although following looser rules of engagement,

Figure 1.7 White Hill Lake Marsh, Virginia, USA. This marsh protects the oysters in Broad Bay and Lynnhaven Inlet on the Chesapeake Bay. The marsh filters water that flows into the inlet. © James Shelton/Marine Photobank.

the UN-Oceans network was subsequently established, although it is clear that the momentum from the Rio Summit to enhance coordination and streamline ocean institutions, according to the programme contained in Chapter 17 of Agenda 21, was partially lost.

I am painfully aware that this is a '*tour de force*' of names of little known agencies, secretariats and a soup of acronyms, only known to insiders, but this is what the UN system offers today. It also offers one too many choices for a forum in which an issue can be discussed. As is well known to international lawyers, the first thing that they advise their clients is to choose the court in which they want their case to be heard. In my view, to recover the momentum, a larger forum with strong participation of civil society is needed: a truly multi-stakeholder forum for the ocean. An honourable first step is the Global Forum of Ocean Coasts and Islands, which the IOC initiated with Biliana Cicin-Sain to promote the Ocean Agenda on our way to the 2002 Johannesburg World Summit on Sustainable Development (WSSD). The Forum succeeded in putting back the ocean in WSSD and has survived thanks to the dedication of a tiny group of

Figure 1.8 Here in Tamil Nadu, India, there are coastal farms for cultivating agar, a gelatinous substance derived from seaweed and used in ice-cream and other foods. © Joerg Boethling/Still Pictures.

friends, leaders and donors, but it needs now to tackle the challenge of its internal governance to reach new levels of service to the whole world community.

Surveillance, enforcement and the status quo

Today this special Ocean international space is regulated by 589 bilateral and multilateral agreements, a fact that is in itself an indictment, reflecting the low priority that the demand for improving environmental or ocean governance is accorded among world political leaders. It also reflects the general complacency of users of this special international space with the continuance of the 'status quo', regardless of its ineffectiveness as a regulatory framework. Furthermore, as the current piracy crisis has revealed, many dangerous gaps exist in the governance of this unique international space. Enforcement of international agreements is usually the responsibility of each state party. In UNCLOS this responsibility is exerted by coastal states, flag-states and port-states. There is a wide range of good and bad practices that could be catalogued for each of the major conventions and agreements. For example, the practice of using a flag of convenience in shipping is widely accepted as a lesser evil and forms part of the status quo.

On the other hand, finding solutions is not easy, as the Somali piracy crisis has shockingly showed the world. Expanding the role of the defence community for 'constabulary' and 'benign' roles in the coast always faces cultural and practical obstacles. Even in the case of prosecuting criminal acts, restrictions exist on most navy organizations to undertake 'police functions' in terms of law enforcement outside national territorial waters. When such actions are possible, the tangle of legal arrangements necessary to render it effective soon becomes overwhelming. In the Caribbean region, for example, with an intensive illegal traffic of drugs and people, the US Coast Guard works under more than 22 bilateral agreements, allowing for law enforcement within the territorial waters of other countries.

But there are positive signs. The Malacca Strait is a critical and strategic waterway in the global trading system. It carries more than one-fourth of the world's commerce and half the world's oil. In 2006, having rejected a previous offer of the USA to patrol the strait, Singapore, Malaysia and Indonesia signed the Straits of Malacca Patrol Joint Coordination Committee Terms of Reference and the Standard Operation Procedures to act jointly in order to tighten security in the Strait. In 2008 Thailand too became part of the joint committee for joint air and surface patrols. This combined effort had a decisive effect in limiting the piracy activity in the strait. In 2004 there were 38 cases of piracy, just two in 2008 and two in 2009.

The bare minimum: the oceans under permanent review

In a pure and simple application of the precautionary approach, because of the alarming trends described above, the World Summit on Sustainable Development in 2002 decided to keep the oceans under permanent review via global and integrated assessments of the state of ocean processes. This conclusion was reached because there were worrying signs that the sector by sector arrangements to manage ocean activities had proven ineffective. Activities were being assessed and regulated independently: there was no instrument to identify the combined impact of all the separate activities on the health of the ocean. The same year, the UN General Assembly not only endorsed the outcome of WSSD, welcoming the Johannesburg Plan of Implementation, but also, and I quote from paragraph 45 of Resolution 57/141, '*Decides to establish by 2004 a regular process under the United Nations for the global reporting and assessment of the state of the marine environment, including socio-economic aspects, both current and foreseeable, building on existing regional assessments*'. This is the most comprehensive initiative undertaken by the UN system yet to improve Ocean Governance.

Figure 1.9 Even large cargo ships and tankers can be boarded by pirates.
© Fedor Salivanov/Shutterstock.com.

Not surprisingly, the initiative faced some resistance in the General Assembly. Countries not sufficiently attentive to these developments adopted a non-committal attitude. Some others questioned the very reasons why an integrated process was being developed, defending the view, for example, that living marine resources should be excluded from the exercise because the FAO had 'exclusive jurisdiction' to pass judgement on fisheries issues. After a couple of intergovernmental Workshops, in 2005 the UN General Assembly finally managed to move forward through resolution 60/30, requesting UNESCO's Intergovernmental Oceanographic Commission (IOC) and the United Nations Environment Programme (UNEP) to take the lead in getting the process started.

The report of 3 years of work is impressive. The group of independent experts analyzed over 471 ocean assessments out of a data base of more than 928 assessments and data collections gathered. The study was peer-reviewed by 34 experts, 15 institutions and 29 governments, and since April 2009 has been available in full on the Internet (http://www.unga-regular-process.org/). Following best practices, the Group of Experts fully documented the peer-review process, cataloguing all the questions raised by reviewers and member states, recording if the comment was accepted or not by experts and giving the reason why. A total of 1270 individual comments were addressed by the independent experts.

The report gave an unqualified 'yes' to the question of feasibility and proposed a clear way forward for the Regular Process. By putting together the first ever comprehensive overview of the ocean assessment landscape, the report gives the essential elements to plan and conduct the first Global Integrated Assessment of the Ocean by 2014–15, in full accordance with UN General Assembly resolution 57/141, only 10 years later. The Summary for Policy Makers, translated into the six UN languages, was analyzed at the 63th General Assembly of the UN by a *Working*

Group of the whole composed by all members of the UN in August–September 2009. A large plurality of participating countries agreed that there is an urgent need to conduct an integrated global assessment of the ocean, including socio-economic aspects, and that the main elements of the proposal emerging from the work done by the Group of Experts was a necessary and sufficient basis to start the first cycle. A first cycle of the regular process could be conducted between 2010 and 2015 and at the same time report to the Commission on Sustainable Development (CSD), if the Commission so agrees.

The report proposes a process that can genuinely integrate the different existing institutions, at the global, regional and national level, in order to produce the assessment. Experts insisted that instead of giving first priority to building a full new set of perfect data, it is more relevant to concentrate on applying the best practices and information flowing from the many assessments analyzed in the study and to organize as soon as possible the first cycle. The aim is to build a robust global institutional base that, by using the same principles, can come up with the first integrated assessment of the world ocean. This first assessment, of necessity, will be incomplete; despite being global, it will be uneven in coverage. All the regional organizations that need to participate are not equal and do not receive the same level of support from their respective members. The enhancement of capacity to conduct assessments in some regions is an essential prerequisite that needs to be addressed for the full realization of the Regular Process.

Nations of the world spend and will continue to spend significant amounts of resources in assessments, because they need evidence-based policies for many ocean issues. New activities are being planned, for example, the assessment planned for the next 5 years by the nations surrounding the North Pacific, grouped under the North Pacific Marine Science Organization (PICES), or the next assessments planned under the Oslo and Paris Convention (OSPAR), The Helsinki Commission (HELCOM) and the Mediterranean Plan of Action.

The Commission on Sustainable Development (CSD) has tentatively scheduled a review of oceans and coastal issues in 2014–2015. Beyond the integrated assessment product that is anticipated by 2014–2015, the way in which the assessment process is proposed to unfold is extremely important. The group of experts devised an incremental approach that will be inclusive and influential. The Regular Process is envisaged as a global mechanism/forum which will become increasingly relevant to a range of existing processes and institutions and which will develop focused interim products ('thematic assessments') and benefits to Member States and ocean entities. Implementing the first Integrated Assessment of the ocean is a bare minimum.

After reading this paper and getting the distinct impression that we are still tilting at windmills, I cannot refrain from narrating what until now has been a

Figure 1.10 The Declaration of the 1992 Summit on Sustainable Development in Rio de Janeiro included a special chapter on oceans. Expectations, though initially high, have not been as easily realized in practice. © Paul Springett/Still Pictures.

private conversation I had the privilege to hold in Reykjavík with H. E. Ólafur Ragnar Grímsson, President of Iceland in 2002. Answering one of his questions, I was carefully choosing my words to report on the progress on ocean governance in the UN. He interrupted my polite discourse to say: 'Why are you being so careful, young man? When in fact we all can recognize now the big mistake we made in 1945 by not creating the UN Ocean Agency. If there's one area where we need the UN most, it's the ocean'. Now, talking of creating a new UN agency is perhaps the most politically incorrect statement you can make in the UN, a sort of political suicide, but I am a firm believer that unless we think out of the box, as President Grímsson was doing that afternoon in Reykjavík, we will fail to stand up to our responsibilities *vis a vis* the ocean and future generations.

2 The United Nations, oceans governance and science

ALAN SIMCOCK

Alan Simcock has been a senior UK marine-policy administrator and Chairman (1996–2000) and Executive Secretary (2001–2006) of the OSPAR Commission for the North-East Atlantic. He was co-chairperson of the first three meetings of the UN informal consultative process on the oceans.

This chapter aims to discuss the structures that have been set up at the world level to manage human impacts on the oceans and seas, the pressures that led to them and the needs for better coordination and scientific support. It is based on a keynote speech given at the Ministerial Round Table on the Oceans, held in 2009, in conjunction with the 35th General Conference of the United Nations Educational, Scientific and Cultural Organization.

Towards UNCED

In the late 1960s and early 1970s, a series of high-profile maritime accidents and events drew attention to the need for a better framework for managing the way in which the world interacts with its oceans and seas. In March 1967, the wreck of the 120 000-ton tanker *Torrey Canyon* on the Seven Stones off south-west England showed how vulnerable coastal ecosystems were to accidents, and how limited international rules were on dealing with them. Concerns about pollution, focused by the 1962 book *Silent Spring*, were brought into the marine context by the *Stella Maris*, which in September 1971 tried (unsuccessfully) to dump 650 tonnes of toxic chemicals in the North Sea. In 1972 a 'Cod War' broke out between Iceland and Germany and the United Kingdom over Icelandic claims to a larger fisheries jurisdiction.

Figure 2.1 The Informal Consultative Process in session at the United Nations, 2006. Image by David Pugh.

The 1972 Stockholm Conference on the Human Environment laid down important principles on the duties of States towards the environment beyond their jurisdictions. Among other developments that flowed from Stockholm was the process that led to the London Convention on Ocean Dumping and the start of the still-developing IMO series of conventions on shipping. The newly created UN Environment Programme started setting up the Regional Seas Programmes around the world, alongside the separate conventions that emerged in 1972 and 1974 to coordinate international action in the north-east Atlantic and the Baltic to protect the marine environment. Many of the Regional Seas Programmes have resulted in similar regional conventions.

In November 1967, the Maltese ambassador to the United Nations had drawn attention to the problems of the seas and called for 'an effective international regime over the seabed and the ocean floor beyond a clearly defined national jurisdiction'. This led to the Third UN Conference on the Law of the Sea. The outcome of this was the agreement in 1982 at Montego Bay in Jamaica of the UN Convention on the Law of the Sea (UNCLOS).

All this marked the start of a new era in international action on the oceans. UNCLOS provided for the first time a unified, coherent framework for the world's oceans and, therefore, required States and others involved in managing human activities that impact on the oceans to start to think in a more integrated way about their actions. Among many other things, it set out the duties of States to cooperate to protect and preserve the marine environment and to cooperate in the conservation and management of the living marine resources of the high seas.

UNCLOS is noteworthy in that, unlike many international conventions concerned with policies that are bound to develop, it contains no provisions for meetings of the States Parties to consider issues generally. It provides for the UN Secretary-General to summon meetings of the States Parties, but, to the extent that it is explicit, specifies only limited, essentially administrative tasks for those meetings. This is largely because many of those negotiating the Convention intended the UN General Assembly to fulfil the role of the general forum for UNCLOS issues. This expectation was largely fulfilled by the way in which the General Assembly took up the task of pursuing the entry into force of the Convention, and in 1996 (when that had been achieved) converted the annual agenda item on UNCLOS into a more general item on the oceans and the law of the sea.

UNCED and its follow-up

The twentieth anniversary of the Stockholm Conference led to the United Nations agreeing to hold the UN Conference on Environment and Development (UNCED) – the 'Earth Summit', in Rio de Janeiro, Brazil in 1992. Before UNCLOS had come into force, but when its provisions (apart from Part XI) had widely come to be regarded as stating customary international law, the members of the preparatory committee for UNCED turned their minds to what needed to be done to promote and support such cooperation.

The outcome was one of the chapters of Agenda 21, the Programme of Action for Sustainable Development, which was adopted (along with the Rio Principles and the Convention on Biological Diversity) by UNCED. Chapter 17 dealt with the 'Protection of the Oceans, All Kinds of Seas, Including Enclosed and Semi-enclosed Seas and Coastal Areas and the Protection, Rational Use and Development of Their Living Resources'. It recommended seven programme areas of which the first was integrated management.

This reflected a new recognition of the need for the integration of the different aspects of management. The text recorded that 'Despite national, subregional, regional and global efforts, current approaches to the management of marine and coastal resources have not always proved capable of achieving sustainable development, and coastal resources and the coastal environment are being rapidly degraded and eroded in many parts of the world'. It further recorded the commitment of coastal States 'to integrated management and sustainable development of coastal areas and the marine environment under their national jurisdiction', and recommended various actions to deliver on this commitment.

Regarding the international level, it stated that 'The role of international cooperation and coordination on a bilateral basis and, where applicable, within a subregional, interregional, regional or global framework, is to support and supplement national efforts of coastal States to promote integrated management and sustainable development of coastal and marine areas'. Chapter 17 went on to state that 'States commit themselves, in accordance with their policies, priorities and resources, to promote institutional arrangements necessary to support the

Figure 2.2 On the UN terrace in New York. The list of acronyms for the various Intergovernmental Agencies with marine responsibilities makes a complicated cocktail mix. Image by David Pugh.

implementation of the programme areas in this chapter (paragraph 17.116)'. Further paragraphs addressed management-related activities at the global level and made specific recommendations that 'The General Assembly should provide for regular coordination, within the United Nations system, at the intergovernmental level of general marine and coastal issues, including environment and development matters, ...'. It goes on to enumerate steps that the UN Secretary-General and the executive heads of United Nations agencies and organizations should take to improve coordination on these issues.

One of the other recommendations of Chapter 17 was on the need for better protection of the marine environment from land-based sources of pollution. Agenda 21 invited the UNEP Governing Council 'to convene, as soon as practicable, an intergovernmental meeting on protection of the marine environment from land-based activities'.

A series of intergovernmental meetings culminated in a conference in Washington DC, United States of America, in October 1995. This conference adopted the Global Programme of Action on the Protection of the Marine Environment from Land-Based Activities. As the intergovernmental meetings worked through the many source categories which impact on the marine environment, it became clear both that many international agencies needed to cooperate in implementing the Global Plan of Action, and that there was no clear forum where such cooperation could be coordinated. This lesson was re-emphasized as efforts were made to get the various international bodies to agree to play their parts in the implementation of the Global Programme of Action – the same issues had to be debated over and over again in the different bodies, and it was extremely difficult to maintain any consistent approach.

The concept of 'coordination' is crucial to effective cooperation between agencies. Coordination can be achieved in many ways, but two different approaches must be distinguished:

external coordination or 'top-down' coordination, when one body sets a binding framework for the actions of other bodies;
peer coordination or coordination between equals, when a group of bodies agree between themselves how to align their actions.

In the international context, where States are equal, and international organizations are controlled by their member States, only the latter is feasible. Agenda 21 called for coordination of this kind within States, among States and among international organizations. Improvements in cooperation and coordination therefore needed – and needs – to be considered at these two different levels.

A global forum for States

In the 1990s, the Commission on Sustainable Development (CSD) began its work. The CSD was set up as a high-level body, technically a functional Commission of the UN Economic and Social Council (ECOSOC). Its main goals are to 'enhance international cooperation and rationalize the intergovernmental decision-making capacity for the integration of environment and development issues', and to examine progress in the implementation of Agenda 21. Under the programme of work which the CSD adopted at the outset, the first review of Chapter 17 (Oceans and All Seas) of Agenda 21 was held at the fourth meeting of the CSD in 1996.

In preparation for this review, the Governments of Brazil and the UK organized in December 1995 a first London Workshop on Environmental Science, Comprehensiveness and Consistency in Global Decisions on Oceans. This was attended by a wide range of countries, and chaired by Ministers from Brazil and the UK. At this workshop, general agreement was reached on 'the need to improve the effectiveness of the means for providing the scientific advice needed both for the formulation of priorities for global action and for ensuring a consistent base for action between the various agencies involved'.

The 1996 session of the CSD (CSD-4) considered the need for international coordination and cooperation on the oceans. It concluded that there was 'a need for a periodic overall review by [CSD] of all aspects of the marine environment and its related issues as described in Chapter 17 of Agenda 21'. But no agreement was reached on how better coordination could be achieved on a more regular basis.

The programme of the CSD ensured that there was a further review of Chapter 17 and oceans at the seventh meeting of the CSD in 1999 (CSD-7). In preparation for this, the Governments of Brazil and the United Kingdom organized a second London Oceans Workshop. Representatives of 39 States from all parts of the world took part, together with a range of international organizations and non-governmental organizations. This workshop concluded that, 'because of the number of different international organizations involved, improved arrangements are needed for co-ordination in order to produce an integrated overview and generate consistent approaches in the various forums to the conservation and sustainable use of the seas'.

This then was the background against which in 1999 the Ad Hoc Intersessional Working Group established to prepare this aspect of CSD-7 debated the question of improving international cooperation and coordination on the oceans. Agreement was reached on the need for improvements, on the fundamental role of the Secretary-General's annual report, on the need to improve AAC/SOCA (see below), and on the need to achieve a better basis for the annual debates on the oceans in the UN General Assembly. Various ideas were put forward on how this might be done (ranging from a Committee of the Whole of the General Assembly to a high-level symposium), but no consensus emerged. Building on these discussions, CSD-7 was eventually able to reach consensus on recommendations to ECOSOC and the UN General Assembly that there should be an open-ended informal consultative process on the oceans – a recommendation that was essentially adopted and given force by UN General Assembly Resolution 54/33.

The basis of CSD-7's recommendation was its conclusion that 'oceans and seas present a special case as regards the need for international coordination and cooperation' and that, building on existing arrangements, a more integrated approach is required to all legal, economic, social and environmental aspects of the oceans and seas, both at intergovernmental and interagency levels.

The General Assembly Resolution formulates the objective for the Informal Consultative Process (ICP) of providing such an approach in terms of 'facilitat[ing] the annual review by the General Assembly, in an effective and constructive manner, of developments in ocean affairs by considering the Secretary-General's report on oceans and the law of the sea'. The Secretary-General's Report is a magisterial overview of the past year's developments affecting the oceans. The expectation was that this would draw together the issues on which coordination was needed, and thus form the basis of a sound debate on the way forward. Three strands can be identified within this objective:

to enable oceans issues to be considered in the round, rather than from one or other sectoral viewpoint;

to deepen understanding of oceans issues among those involved in negotiations in New York, who may well not be oceans experts;

to permit a dialogue between the UN Member States and those charged with coordination among the UN Secretariat, the specialized agencies and other bodies.

Abb. 335. Der Ekmansche Repetierstrommesser für Serienmessungen.

Figure 2.3 Marine scientific expeditions have gradually built up a picture of the oceans and how they behave, often using very simple instruments in the early days. Here a basic current meter is used on the 1925–1927 Atlantic German Meteor expedition. © NOAA.

What are the most significant elements of the ICP mandate? There seem to be five:

(1) The General Assembly resolution stresses that everything is to be consistent with UNCLOS.

(2) There is the three-fold description of the process:

open-ended – open to all UN States, and special arrangements were made to promote the participation of developing countries and international non-governmental organizations from those countries;

informal – no special organizational arrangements were made: the only officers are the two co-chairpersons, and the secretariat is provided by DOALOS. Since it was informal, arrangements could be made for international organizations and international non-governmental organizations to participate; and

consultative – no powers of decision or management.

(3) It is to take account of the differing characteristics and needs of the different regions of the world, and is not to pursue legal or juridical coordination among the different legal instruments – another aspect of its informal nature.

(4) The process may propose elements for the consideration of the General Assembly in relation to Assembly resolutions under the agenda item entitled 'Oceans and the law of the sea'. This is the core of the process. By focusing on what it might be appropriate for the General Assembly to consider, the process was enabled both to give a new edge to the interventions of the General Assembly on coordination questions, and to enhance the Assembly's negotiations and debates on these issues. The outputs of the process were therefore twofold – material on issues for consideration by the

Assembly and reports of the discussions during the meetings of the process, which provided a summary of information relevant to those issues and a note of the points relevant to them which were of concern to the participants.

(5) Like all new United Nations structures, it has a 'sunset clause' – after 3 years the effectiveness and utility of the process was to be reviewed. The ICP was confirmed in 2003 and 2006 and devoted its 10th session in 2009 to focusing on the implementation of its outcomes, including a review of its achievements and shortcomings in its first nine meetings. The Eleventh Session will be held in June 2010.

Inter-agency coordination

As has been said, Agenda 21 called for better coordination within the UN family, as the second part of improved cooperation and coordination at the global level. To respond to this, the UN's Administrative Committee on Coordination (ACC), which brings together the Secretary-General and the heads of the specialized agencies, programmes and similar bodies, set up in 1993 a Sub-Committee on Oceans and Coastal Areas (SOCA), with the remit of coordinating the follow-up of Agenda 21, and the reports to the UN bodies on its implementation. This body, however, provided no means for establishing a dialogue between the agencies and the UN Member States. In November 2001, moreover, its parent body (the Administrative Committee on Coordination) decided to abolish all its existing subsidiary bodies and instead rely on ad hoc, time-bound, task-oriented arrangements.

There was considerable concern among many States at the disappearance of SOCA, which (although not highly effective – the CSD had called for improvement as early as 1996) had at least provided a forum for interagency coordination. As a result of recommendations from the ICP and the decisions of the 2002 Johannesburg World Summit on Sustainable Development (WSSD), the 57th General Assembly in 2005 invited the UN Secretary-General to establish an effective, transparent and regular interagency coordination mechanism on oceans and coastal issues.

The result has been UN-Oceans (which emerged less formally after the abolition of SOCA as the Oceans and Coastal Areas Network). This brings together not only UN specialized agencies and programmes, but other relevant entities and international conventions. The full roll-call of those who could take part includes 23 organizations (see the annex for a full list) and ranges from Divisions of the UN Secretariat through specialized agencies to the secretariats of international conventions. UN-Oceans is headed by a coordinator and deputy coordinator, and its secretariat is provided by UN-DOALOS (which is the organizing secretariat and

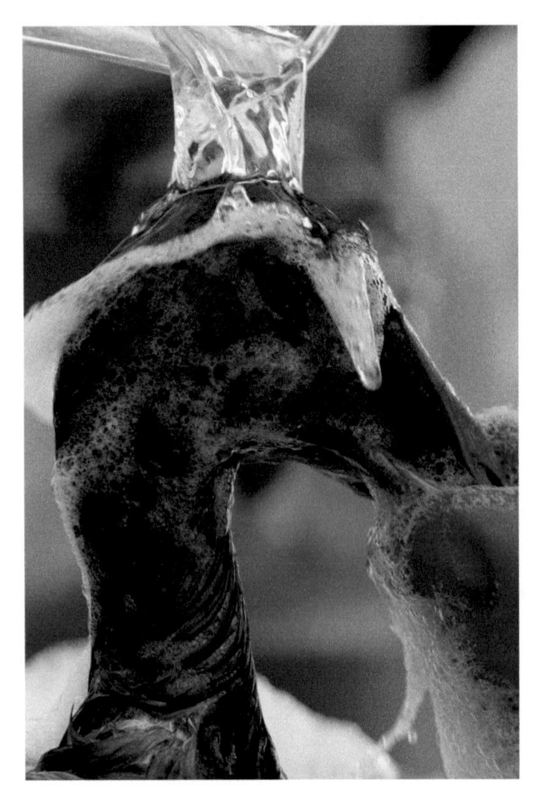

Figure 2.4 A duck is cleaned by volunteers after Norway's most catastrophic oil spill in 2009. © Linda Schonknecht/Marine Photobank.

thus organizes meetings and tele-conferences and produces their reports) and UNESCO-IOC (which is the implementing secretariat and thus maintains information about relevant programmes and activities, reviews this information and draws to the attention of the partners gaps, duplications or potential for collaboration and to follow up actions developed by UN-Oceans that are not clearly the responsibility of anyone else).

The very length of this list shows how many fields of human (and therefore governmental) activity touch on the oceans. This multiplicity of interests makes it clear that it would not be possible for any one organization to cover the whole field, and is therefore a strong argument against any attempt to create a new oceans government mechanism. The problem is how better to coordinate action at all levels.

The input of science

In all fields, the whole cycle of policy making and implementation has to begin, and end, with collecting and analyzing information about what is happening: the initial collection and analysis leads to policy formulation, to implementation, to monitoring and to evaluation, which feeds back afresh into policy review and development. Equally, efficient and effective coordination must be based on good information. The oceans are no different. This has been recognized by the ICP and the WSSD, which have both called for a mechanism for the global reporting and assessment of the oceans.

In 2005, the UN General Assembly initiated such a mechanism by setting up the first stage of the Global Reporting and Assessment of the Marine Environment (GRAME), in the form of an 'assessment of the assessments' that have been made globally, regionally and (where relevant) nationally of the world's oceans and seas. The Group of Experts for the assessment of assessments has now reported. At the end of August 2009, an open-ended workshop of the General Assembly started considering their proposals on how the regular process of global reporting and assessment can be managed. We await the General Assembly's conclusions on how the process can be carried forward.

The proposals of the Group of Experts would provide a regular system for keeping the world's oceans under review. It would offer an integrated assessment which would look at environmental, economic and social aspects together – something that we do not yet fully have anywhere. This factual basis on what is happening in the oceans and how they may be expected to develop is a fundamental tool for improving management. As the First London Oceans Workshop concluded, we need better scientific advice both for the formulation of priorities for global action and for ensuring a consistent base for action among the various agencies involved.

Conclusion

For managing human impacts on the oceans and seas, we have to consider the national, regional and global levels. If we look at these levels, we see that, since the first, formal, global agreement on the need for integration in 1992:

at the ***national level***, many States have improved their systems, with some creating new ministries or directorates to bring together aspects of maritime policy or to coordinate it;

at the ***regional level***, there are more regional agreements, and existing ones have been updated and strengthened; and

at the ***global level***, we now have a forum for States to look at oceans issues in the round, and to underpin the work of the UN General Assembly. We have an interagency coordination mechanism. And we have proposals for Global Regular Process for Reporting and Assessing the Marine Environment.

What more do we need? Three main needs can be identified:

(1) We need to ensure that the gains of the last two decades are maintained – if commitment to them fades, the threats to the oceans and seas will worsen, because coordination will weaken. The commitment must come both from States and from the permanent staff of the intergovernmental organizations. Consistency at the national level is needed, in particular that the States' representatives follow a consistent policy in the various international organizations.

(2) We need to ensure that an effective, regular Global Reporting and Assessment process is put in place to underpin the coordination processes that have been

Figure 2.5 Oil spillage washed up on Stoke beach in 1990 after tanker and trawler collide off the Devon Coast, UK. © David Cayless/Marine Photobank.

established. Without first-class science, the debates at the various levels may miss the important issues or get priorities wrong.

(3) We need to find ways of binding together the different coordination processes. If they continue in isolation, significant opportunities will be lost. UN-Oceans and the ICP need to have a real dialogue on the main issues. The regular, global assessments need to be taken up by UN-Oceans and the ICP, and consideration given to what the assessments show, focused on action to meet the challenges that emerge.

How can this be achieved? One essential is for States to implement their commitments in Agenda 21. A second requirement is for States and international organizations to provide sufficient resources for the Informal Consultative Process and UN-Oceans to fulfil their roles – in particular,

Figure 2.6 Marine shells form a complicated beach pattern, as diverse as the range of international bodies concerned with the sea. © Tadija/Shutterstock.com.

resources to permit developing countries and their experts to participate fully. To help these mechanisms to function well, however, a link between them is needed. The UN General Assembly has already noted the important role of the annual report of the UN Secretary-General on oceans and the law of the sea. This is already a magisterial summary of what ***has*** happened over the past year in this field. It could be developed so that, on this factual basis and on the basis of the Global Process of Reporting and Assessment, it highlights the crucial issues that the General Assembly needs to address, or to ensure are addressed by the international community through one or other of its mechanisms.

Annex

Organizations that can participate in UN Oceans:

Food and Agriculture Organization of the United Nations (FAO)
International Atomic Energy Agency (IAEA)
International Bank for Reconstruction and Development (World Bank – IBRD)
International Hydrographic Organization (IHO)
International Labour Organization (ILO)
International Maritime Organization (IMO)
International Seabed Authority (ISA)
Organization for Economic Cooperation and Development (OECD)
Secretariat of the Convention on Biological Diversity (CBD)
Secretariat of the Ramsar Convention on Wetlands (Ramsar)
Secretariat of the United Nations Framework Convention on Climate Change (UNFCCC)

UN Secretariat – Department of Economic and Social Affairs (UN-DESA)
UN Secretariat – Division of Ocean Affairs and Law of the Sea (UN-DOALOS)
United Nations Conference on Trade and Development (UNCTAD)
United Nations Development Programme (UNDP)
United Nations Educational, Scientific and Cultural Organization (UNESCO), and its Intergovernmental Oceanographic Commission (UNESCO – IOC)
United Nations Environmental Programme (UNEP)
United Nations Human Settlements Programme (UNHSP – UN-HABITAT)
United Nations Industrial Development Organization (UNIDO)
United Nations University (UNU)
World Health Organization (WHO)
World Meteorological Organization (WMO)
World Tourist Organization (WTO)

3 Marine scientific research and the United Nations Convention on the Law of the Sea

ELIE JARMACHE

Elie Jarmache is an expert on marine law. He has been Head of the French delegation to the United Nations Commission on Limits of the Continental Shelf, and is the founding Chairman of the Advisory Body of Experts on the law of the sea of the Intergovernmental Oceanographic Commission (IOC/UNESCO). At present he is in the French government Secrétariat Général de la Mer responsible for maritime law, biodiversity, continental shelves and scientific research.

The United Nations Convention on the Law of the Sea (UNCLOS) entered into force on 16 November 1994. It defines the rights and responsibilities of nations in their use of the world's oceans, establishing guidelines for businesses, the environment and the management of marine natural resources. In addition to more general scientific references, Part XIII of UNCLOS deals specifically with Marine Scientific Research, and while encouraging States to promote and facilitate research, it also declares that States have the right to regulate, authorize and conduct marine scientific research in their exclusive economic zone and on their continental shelf. Marine scientific research projects, undertaken by or under the auspices of international organization, are addressed in Article 247. As one of the principal international organizations concerned with marine scientific research, cited in UNCLOS, the Intergovernmental Oceanographic Commission of UNESCO has regularly addressed UNCLOS and its implications.

When it was established 50 years ago in 1960, the Intergovernmental Oceanographic Commission (IOC) fell within the legal framework of the four Geneva Conventions on the Law of the Sea of 29 April 1958. Those conventions covered various fields, namely the continental shelf, the high seas, the territorial sea, fishing and conservation of the living resources of the high seas. That framework was

Figure 3.1 The United Nations headquarters building in New York is the venue for meetings of the State Parties to UNCLOS. Image by David Pugh.

characterized by a fragmented vision of maritime and marine governance, and was also very limited in nature because very little provision was made for marine scientific research. Only the Convention on the Continental Shelf refers to marine scientific research explicitly in a few lines, in paragraphs 6 and 8 of Article 5. This is undeniably very little for an area that will develop in step with growth and the need for knowledge and with advancements in science and technology.

The IOC was founded to meet the need for an institutional and international cooperation mechanism in the field of marine sciences, even though its existence was tied to that of UNESCO, whose fields of competence have, from the outset, included science. Conceivably, the founders of the IOC wished to highlight the specificity of oceans and of scientific research relating thereto. However, IOC pioneers did not make the IOC fully independent. Caution led them to place the new organization within UNESCO, and that duality subsists 50 years on: the IOC is both a UNESCO programme and an intergovernmental organization. The optimist may analyze such a duality as a permanent miracle, but it may also be regarded as a source of ambiguity, giving rise to the IOC's malaise that surfaces whenever it embarks on self-evaluation, or debate on its future.

Despite this unusual situation in the sphere of international organizations, the quality and innovativeness of the programmes, it was generally considered that the IOC did indeed meet the international community's needs. The IOC's ability to give heed to the States that acceded to independence in the late 1950s and early 1960s, when the IOC was being established, helped to strengthen its legitimacy. This holds true to this very day; the Latin American, African and Asian States are still some of the most faithful members of the Organization and, concurrently, beneficiaries of IOC activities.

Figure 3.2 UNCLOS has special conditions for the operation of Research Ships in the EEZ of coastal states. © Leighton Rolley, National Oceanography Centre, Southampton.

The IOC has therefore existed for several decades in a relatively stable legal setting with regard to the supervision of marine scientific activity. It must be pointed out that the international law of the sea acknowledged only two sea areas: the territorial sea and the high seas. As the breadth of the territorial sea had not been set at the time, there was uncertainty as to the seaward extent of a State's sovereignty. Under customary law, the accepted practice was a breadth of 3 or 6 nautical miles. Another consequence was that the tenet of the freedom of high seas was becoming a dominant principle under which marine scientific research was established. The situation could be summed up as follows: no maritime area between the territorial sea and the high seas was under national jurisdiction, and scientific research was by nature unrestricted. To complete the picture, it is important to note that the potential and capacities for marine sciences were largely concentrated in a few States in the Northern hemisphere that were traditional naval powers. Those States had long acquired the culture of the freedom of the

Figure 3.3 Rough seas at Pedra Branca Rock, off the south coast of Tasmania, Australia. Image courtesy of Peter Dexter.

high seas that applied to their activities beyond the territorial sea, for which no set limit existed in international law inasmuch as it varied from 3 to 6 nautical miles and occasionally up to 12 miles. This situation persisted owing to the failure of the 1960 Conference on the Law of the Sea to establish the breadth of the territorial sea.

The period of calm for Researcher States and their scientific communities was broken by the Third United Nations Conference on the Law of the Sea of 1974, which led to a ground-breaking convention in April 1982 – the United Nations Convention on the Law of the Sea, well known by its English acronym UNCLOS.

Under the Convention, new maritime areas were established, namely the Exclusive Economic Zone (EEZ) and the seabed and subsoil recognized as the Area and placed under the responsibility of the International Seabed Authority. The Convention provides that the continental shelf of coastal States may be extended beyond 200 miles, under certain conditions and subject to the approval of the Commission on the Limits of the Continental Shelf, but may not exceed 350 nautical miles.

The old legal order governing seas and oceans was challenged on political and economic grounds. Developing countries wished to assert their rights over natural resources, particularly the living resources, beyond their territorial seas. These States lacked the means to gain access to those resources, which they considered to be a source of income. The 1970s were conducive to such claims because the doctrine of the new world economic order was then prevalent. The compromise reached was the EEZ and rights over natural resources were qualified as sovereign.

The new zone, which also comprised the continental shelf, was marked by an innovation in that marine scientific research was placed under the jurisdiction of coastal States. The IOC must live with this new rule. It was introduced more than a quarter of a century ago, yet, the IOC and some of its Member States have not totally adapted.

The Convention on the Law of the Sea has changed the existence of the IOC because it has changed the life of coastal States and Researcher States Members of the IOC.

Marine scientific research, the IOC's core mission, is now an unrestricted activity only on high seas and, to some extent, on the seabed area beyond national jurisdiction. Coastal States have extended their jurisdiction beyond their territorial seas in order to authorize and regulate marine research for scientific purposes. The purpose of Part XIII of UNCLOS is to establish the rules governing scientific research, which rests on the principle of the consent of the coastal State, such consent being expressed in respect of the territorial sea or implied in respect of areas under national jurisdiction, pursuant to Articles 245 and 252 respectively. Without reviewing these provisions in detail, it is noteworthy that the first two sections of Part XIII legitimize a *posteriori* IOC's establishment and activities.

Section 1 establishes the general principles for marine scientific research, stating that competent international organizations have the right to conduct research (Article 238). The IOC was established in part for that purpose, and UNCLOS arguably confirms the terms of reference adopted 50 years ago, in 1960. The second lesson to be learned from the Convention is contained in the Article 240 list of guiding principles applicable to scientific research. The first two principles can be of great importance today and are worthy of mention – that 'marine scientific research shall be conducted exclusively for peaceful reasons', and it 'shall be conducted with appropriate scientific methods and means compatible with the Convention'.

Section 2, entitled 'International Cooperation' is also highly important in view of the very nature of the IOC. This intergovernmental body rests on the mechanism of cooperation among States and, to a greater extent, among marine scientific research stakeholders. Cooperation, outlined in Section 2, has but one goal: the publication and dissemination of information and knowledge, as clearly stated in Article 244. However, cooperation is not possible if the favourable conditions, guaranteed by competent international organizations – of which the IOC is, to some extent, the leader in the field – are not created. It is therefore important to stress the role of the IOC, which brings communities of researchers, such as ICSU and SCOR, together.

The IOC carried out these functions naturally before UNCLOS was implemented. However, the inclusion of non-researcher developing States in the Convention drafting process has heightened awareness of the importance of marine research activities and of knowledge dissemination in ensuring that all stakeholders benefit.

The Convention on the Law of the Sea thus vindicates the foresight of the founders who established the IOC in 1960. UNCLOS has also been a serious challenge to the IOC. The question now is whether, how and with what results the IOC has risen to the challenge.

The United Nations Convention on the Law of the Sea (UNCLOS) entered into force on 16 November 1994, an important date as the text adopted in 1982 then became an applicable source of law, a reference for States and international organizations, some of which are recognized as being competent. The IOC realized that its activities would take place in a totally overhauled legal framework that it could not ignore. In 1997, the Assembly of the IOC adopted a resolution (XIX-19) that established the Advisory Body of Experts on the Law of the Sea, well known under its acronym ABE-LOS. The Advisory Body was entrusted with providing advice upon request to the IOC governing bodies, namely the Assembly and the Executive Council, as well as to the IOC Executive Secretary on any matter concerning the application of the United Nations Convention on the Law of the Sea by the IOC.

The law of the sea then became an integrated part of the IOC's work, and the revised Statutes of IOC, in particular Article 3, directly referred to the United Nations Convention on the Law of the Sea.

One specific UNCLOS provision, namely Article 247, deserves mention because it took on particular importance for the activities of the IOC. Indeed, the attention paid to Article 247 set in motion all the work leading to the establishment of the Advisory Body of Experts on the Law of the Sea and to the examination of relations with the United Nations Convention on the Law of the Sea. Article 247 certainly constitutes a call for greater cooperation among States through the competent organizations that promote, facilitate and coordinate research projects. The IOC has this role in marine science. As a result of ABE-LOS' thorough work, the IOC succeeded in adopting a 'Procedure for the application of Article 247' by means of Resolution XXIII-8, which was a significant step in the IOC's implementation of the United Nations Convention on the Law of the Sea, and a step without precedent in other international organizations. The same measure was taken with regard to the transfer of marine technology, an area within the IOC's responsibilities.

Part XIV of the United Nations Convention on the Law of the Sea treats this subject. In Article 271, the IOC found the basis for ABE-LOS' activities and was able

Figure 3.4 A view of the General Assembly Hall at the Palais des Nations, during the opening meeting of the Conference on the Law of the Sea, February 1958. © United Nations.

to fulfil its role in the promotion of guidelines, criteria and standards for the transfer of marine technology. Resolution XX-12 adopts these principles within the framework of the IOC's activities and programmes.

Two other achievements of the IOC as regards the law of the sea deserve mention, namely, the introduction of a procedure for the deployment of Argo floats, and the monitoring of IOC Member States' practices in marine scientific research.

The IOC deemed the first issue to be crucially important insomuch as it raised the question of the nature of the activities grouped under so-called 'operational oceanography' and the rules that apply to them. It was becoming urgent for the IOC to respond in some way because operational oceanography, one of the IOC's major programmes and jointly led by the World Meteorological Organization (WMO), was creating disagreement among IOC Member States due to differing interpretations of the United Nations Convention on the Law of the Sea, in particular as to whether Part XIII applied to that activity. Another question concerned which rules and procedures should apply to Argo floats deployed in the high seas that drifted into a State's EEZ.

These questions were submitted to the Advisory Body of Experts on the Law of the Sea, which, after a few working sessions, was able to recommend a procedure to the IOC governing bodies that was a compromise between the consent regime and the freedom to act without regard for the coastal State. Resolution EC/XLI.4, adopted by the IOC Executive Council on 30 June 2008, describes this procedure.

The second issue, equally important, particularly demonstrates the added value offered by the IOC in the implementation of the United Nations Convention on the

Law of the Sea. The Advisory Body of Experts on the Law of the Sea deemed it useful to develop a questionnaire on Member States' practices in the areas of research and the transfer of marine technology. Like any questionnaire, this one was imperfect when first launched, but by effectively analyzing responses, the IOC and its experts were able to improve it after a few years. The questionnaire is a long-term tool that should be kept. Developing Member States have been able to have access to pertinent information about the United Nations Convention on the Law of the Sea, and to request IOC services and the assistance of experts to improve their practices. The results of the questionnaire also proved useful when the United Nations' Division for Ocean Affairs and the Law of the Sea (DOALOS) initiated the updating process of *Marine Scientific Research: A Guide to the Implementation of the Relevant Provisions of the United Nations Convention on the Law of the Sea*, published in 1991. A large number of the experts asked by the Division for Ocean Affairs and the Law of the Sea to participate in the updating process were ABE-LOS members. Cooperation between the two groups was wholehearted and exemplary.

On the eve of the IOC's 50th anniversary, it should be pointed out that much is still happening with regard to the law of the sea. One could even say that the governance of the seas and oceans is entering a new, important phase in which the IOC must be present. The question is whether the IOC will understand that its governing bodies must continue to focus on the United Nations Convention on the Law of the Sea. The main issues in coming years, such as climate change, marine biodiversity and ocean fertilization, will be addressed in taking legal texts and the needs of marine scientific research into consideration. The IOC must take the full measure of the issues at stake in the protection of the marine environment and the new rules that are required. Yet, from some of the IOC governing bodies' debates and some positions taken by delegations, it appears that this aim and this vision are lacking. The IOC does not seem to realize that scientific research and related techniques can be questioned in terms of their impact on the marine environment. It remains to be seen what rule the IOC should apply in order to move in the right direction.

As a set of rules and practices on the governance of the seas and oceans, the law of the sea, and therefore the United Nations Convention on the Law of the Sea, continue to intimidate the IOC, and many delegations take refuge in the routine examination of various IOC programmes. At one of its sessions on the occasion of its 50th anniversary or shortly thereafter, the IOC needs to create the conditions for a debate on its future participation in the implementation of the law of the sea. The IOC must understand that discussions are under way and that projects have been announced that will change the course of maritime regulations over the medium and long term. It must prepare for these changes, as it did when the United Nations Convention on the Law of the Sea entered into force in 1994. Indeed, as early as 1995, it established an ad hoc open-ended working group to examine the implications of the United Nations Convention on the

Figure 3.5 The debate continues half a century later: a wide view of the General Assembly in New York on 4 December 2009 as it discusses a draft resolution on 'Oceans and the law of the sea'. © United Nations.

Law of the Sea on its activities. It then established its Advisory Body of Experts on the Law of the Sea, which brings together one scientific expert and one legal specialist per participating Member State, an original feature existing nowhere else.

The decade coming to an end has demonstrated the remarkable pertinence of IOC's instincts. On its anniversary, when the page is turning, the IOC must find again its intuition, which has enabled it to take significant initiatives and to assume responsibility for ambitious programmes in the competent international organizations. Ocean governance is going to change because of economic and societal challenges that can only be faced through the cooperation of those in marine sciences with various stakeholders who will raise their standards on responsibility, ethics and the protection of the environment. The existing legal framework will itself be confronted with the challenge of adapting or changing. As a result, oceanographic practices may be altered as scientists adapt to the technical, scientific and legal developments of a new era.

4 Fifty years of building national marine skills

JOANNÈS BERQUE AND EHRLICH DESA

Joannès Berque joined the Intergovernmental Oceanographic Commission's capacity development section in 2005, after a PhD in physical oceanography from the Scripps Institute of Oceanography, California. His work at the Commission has focused on technical cooperation for the sustainable and climate-resilient use of coastal resources.

Ehrlich Desa began his career in 1973 as a Research Scientist with the National Institute of Oceanography in India. In 1994, he was appointed as Director. He joined the IOC in 2003 as Chief of the Capacity Development Section; he is now IOC Director of programmes.

The oceans are a global, interconnected commons that need the active involvement of all countries if we are to understand and manage them sustainably, and better predict and prepare for the consequences of global climate change. Coastal waters of developing countries sustain some of our planet's richest ecosystems and biodiversity – a globally significant heritage for future generations, worldwide.

Not all governments have the capacity to conduct marine scientific research and observations and to apply that knowledge effectively. More advanced countries must help by transferring and fostering the necessary skills. The IOC, as the recognized United Nations mechanism for global cooperation in the study of the oceans, carries a natural responsibility in these matters.

Yet, there are limits to the role external actors can play in developing marine scientific capacities in another country. Some countries achieved rapid growth in

Figure 4.1 Sharing coastal space. An artisanal fisherman at work while in the distance a tourist flies by on jet-skis. Coast near Libreville, Gabon. Image by Joannès Berque.

scientific capacities, where external cooperation played a negligible role. Conversely, there are examples where capacities have grown slowly if at all in spite of substantial external technical assistance. The main message is that a government's determination to focus on a particular scientific priority is an essential ingredient far more important than externally driven cooperation. But even with government determination, scientific excellence and relevance to national priorities are not easily achieved. The ingredients necessary for vibrant science organizations are highly variable depending on the scientific discipline, economic and legislative settings, culture, education system and other associated factors.

Technical cooperation in marine sciences has often shown positive results and benefits to all parties. UNESCO/IOC has a natural lead role and responsibility, in key aspects of technical/scientific cooperation, to assist its Member States in moving towards a more sustainable use of coasts and seas. After 50 years, the Commission's historical remit remains especially relevant in today's context of rapidly degrading marine environments and climate change.

The Intergovernmental Oceanographic Data Exchange (IODE) was a founding programme of the IOC with the vision that data would be freely contributed to a global commons, and resulting knowledge freely disseminated. The IODE had a training component inherent in its initial structure. With the need to fill the knowledge gaps of large areas of our ocean, the data exchange programme grew rapidly over the years supported by nations with strong traditions in ocean research and services. Where interest in the ocean touched developing shores, a programme to build capacity through Training, Education and Mutual Assistance (TEMA) was created in 1973; basic elements of TEMA are intrinsic to the way that we approach capacity development today. Yet the vision of informed and wise governance as evidenced by participation of all States remains unfulfilled.

Figure 4.2 A coastal defence appropriate for a coast endowed with a rich equatorial forest? Tree logs used in an attempt to slow down coastal erosion near Libreville, Gabon. Unfortunately, without necessary studies and proper design, hard defences often worsen rather than mitigate erosion. Image by Joannès Berque.

We hold that too much effort went into short-term training for participation in global programmes, instead of into long-term education of local scientists to address national priorities. Examples abound: the 1980s saw research into El Niño and harmful phytoplankton blooms receive enough attention from the developed world to grow into programmes with assured funding and international and intergovernmental bodies to guide their work and growth, namely into HAB (Harmful Algal Blooms, instituted in 1993) and GOOS (Global Ocean Observing Systems instituted in 1991). These programmes were not of immediate interest to the developing world. In our haste to address the growing list of issues, there was time and funds enough to address only the short-term immediate training component of TEMA, neglecting the more long-lasting process focused on education and research. Thus, scientists from the developing world were trained in specific techniques and methods that allowed them to successfully play their part in global programmes, but were not always used much beyond their workshops either because the programmes were not relevant nationally, or a sense of ownership was not created. This is now changing as previously unadopted global programmes are being increasingly owned by states, as governments are now aware of local impacts. Examples include the HAB programme's importance to seafood exporting countries, and the importance of sea-level monitoring to countries threatened by sea-level rise.

TEMA activities, allowing long-impact educational initiatives to be conceived and implemented, have been very effective. Examples of the Education component of TEMA are: (1) the original Training-Through-Research (TTR) programme, begun in 1991, that brings together oceanographic researchers and their students on a well-equipped research vessel to address commercially important marine topics; (2) the Baltic sea programme, begun in 1991, with a novel two-ship programme (a coastal vessel and a catamaran) and the University-at-Sea programme, begun in 2005, providing berths on-board available research vessels to groom and excite students about the oceans; and (3) IOC University Chairs in Marine Science, started in 2000 empowering

Figure 4.3 A joint IHO and IOC mission to Madagascar discusses new charting ideas with local experts. Image courtesy of the International Hydrographic Bureau (IHB).

selected academics to conduct special training programmes through national or regional workshops for students of oceanography. All of these educational programmes have an excellent and positive influence on students, though increasing their geographical coverage and the numbers trained annually remains a major challenge.

In the remainder of this chapter, we showcase capacity development efforts from a variety of environments, and use these lessons to validate a set of *Principles of Engagement* to guide further capacity development efforts in marine science and observations. The new approach is dedicated to the growth of capacities, in areas chosen by recipient institutes, and focuses on the longer term growth of science organizations within countries, independently of the objectives of global programmes.

A few principles of engagement

From this diversity of experiences, some simple *Principles of Engagement* emerge that seem to work in a wide set of conditions. The foremost observation is the striking difference in effectiveness, sustainability of benefits and impact of resources invested between activities with strong country ownership and those primarily driven by external actors. In the last few years our capacity development activities have attracted financial participation from the recipients that at times exceed international contributions, and surprisingly from participants of some of the least developed nations whose institutes sponsored air travel and/or accommodation to attend workshops. By contrast, there are capacity development workshops that do not manage to attract participants without the perks of full sponsorship, and activities that are effectively extinct in the month following the cessation of even long and substantial external support.

The difference in response we believe arises primarily from different levels of ownership, which trace back to relevance to a country's priorities. Ownership and relevance are complex to achieve and evaluate, but some success factors include

Figure 4.4 Planting for mangrove aforestation for coast protection at Mombasa, Kenya. © Joerg Boethling/Still Pictures.

self-determination of project content, self-determination of the use of project resources (financial ownership) and ownership of project expertise. A scorecard approach to quantifying this evaluation at project level has proved effective. In this approach, several aspects of ownership of resources, project content and expertise are assessed. This conclusion is in line with recent developments in international cooperation in fields other than marine sciences. More importantly for the oceanographic community, the scorecard leads to cooperation respectful of partners, their motivations and their potentials.

Capacity development activities have often, despite good intentions, resulted in a weakening of national capabilities. Usually, this very negative aspect occurs when activities are not focused on national priorities. The external funds associated with externally driven programmes attract marine scientists in developing countries away from state-assigned priorities, a temptation hard to resist given the very unequal appeal of external funds and local salaries, among other things. Indeed many a developing

Figure 4.5 The attraction of aquariums is a powerful combination of education and tourism. © A Cotton Photo/Shutterstock.com.

country's marine institutes become local outposts for projects from international partners – effectively, a developing country is subsidizing the research programme of a developed country!

Responsibility for such a situation partly rests with the government ministry responsible for the institute. The government may invest in the building, infrastructure and salaries of the institute staff, but stop short of providing for research funds. In some developing countries, governments encourage and expect scientists to attract international project funding; this is akin to buying a car and allowing it to be driven by someone else to avoid spending on gasoline. This perpetuates a destructive cycle of dependency on external funds, making the institute less relevant to national priorities, and further losing the interest and support from national funding partners. Whether future well-intentioned funding from globally focused science programmes will continue drawing institutes away from national priorities or not may be the key question of technical cooperation in marine sciences in the next decade.

In summary: the road to global science starts with local priorities. The differences typical of local and global priorities in capacity development are summarized in Table 4.1. Improving national capabilities is a priority for most developing countries, and necessary in the context of the sustainable development goals that we need to achieve globally. The two drivers: local priorities, and increased participation in global programmes, are basically the two dimensions along which development activities are framed. Both are necessary, and often complementary. However, our experience, and the history in many countries, shows that science organizations need to have visible applications to local priorities before they can gain the national support necessary to grow into entities effectively contributing to global science programmes.

In the coming years capacity development at the IOC will conform to the United Nations drive of *Delivering as One* in partnership with unified country teams. This should mean that the Commission will only take part in activities that have been previously spelt out in national plans, and which have national ownership.

Table 4.1 **Comparison of global and local priorities in capacity development**

	Global marine science	Local or national marine science
End goal	Include better prediction of seasonal weather and of climate change, or 'curiosity-driven' science	Better management of marine and coastal resources and environment
Driven by whom?	About 10 most scientifically advanced nations	The remaining more than 100 IOC Member States
Tools used	Large computers, large bandwidth, instruments, boats, large budgets	Small computers, small bandwidths, few or no at-sea instruments, few or no boats, small budgets
Demands on institutes	Publications in scientific journals	Reports and recommendations for a specific national priority
Societal benefits	Longer term, often indirect	Must be visible and immediate, on urgent issues such as fisheries, erosion, pollution, economic development of coastal areas
Resulting capacity development priorities include:		
Measurements	Launching and recuperating Argo floats, *in situ* validations of models or satellite sensors	Issue-based measurements (e.g. water quality, erosion)
Remote sensing	Validation of new sensors by *in situ* measurements	Applications of old sensors (surface temperatures, chlorophyll)
Data transmission needs	Continuous collection of data in standard format and provision to global observing systems	Internet bandwidth to download existing data
Modelling	Global or basin-scale	Coastal modelling, bay, creek or lagoon-scale

Figure 4.6 Fish images on Cuban stamps, an excellent way to inform the public of the fascination of the sea. © Igor Golovniov/ Shutterstock.com.

Figure 4.7 Tourism in the backwaters of Kerala state, India. Ecotourism activities using house boats are becoming very popular in the State, while care is taken to prevent deterioration of habitats. © A. Bijukumar/ Marine Photobank.

There has been real and evident progress in the past 50 years. Countries that have developed their marine science capabilities and interest in marine affairs now offer their experience to the remaining member states, still in the majority. For the future, as the global scientific playing field levels out, the two dimensions of capacity development activities at the IOC, locally driven and globally driven, will gradually merge. Capacity development will evolve into knowledge management and the coordination of global cooperation between equal partners, all contributing to the better understanding of our oceans and the preservation of our common marine and coastal heritage. The founding vision of the IOC, with member states equally informed and participating equally in global ocean governance, will be fulfilled.

Three success stories of institutes that have grown into nationally, regionally or internationally recognized science organizations are reported below in the contributions by colleagues from the Acuarió Nacional Cuba, the Kenya Meteorological Department and the National Institute of Oceanography, India.

Capacity development at the National Aquarium of Cuba

Guillermo Garcia Montero

The National Aquarium of Cuba is a not-for-profit institution officially opened on 1 January 1960, almost coincident with the Intergovernmental Oceanographic Commission (IOC) of UNESCO. We work to create, in Cuba, marine-environment awareness, from the highest level of leaders to the general citizen.

Since its creation, 'capacity development' has been at the centre of our vision and mission objectives. It is our belief that the best way to develop our capacity is through 'doing things'. We consider capacity development as a process. Any activity in consonance with the general goals, objectives and purposes of the institutional mission and vision can contribute. Capacity development at the National Aquarium of Cuba goes much beyond the creation of a school, increasing the numbers of PhDs in marine sciences, or the creation and construction of science institutions.

During its 50 years of work, the National Aquarium of Cuba has contributed to the conception, promotion, coordination, implementation, execution and evaluation of activities designed for the general development of its human, institutional resources as well as the general public. Its main field of work of marine science and environmental education for the benefit of Cubans in general has progressively increased over the years. The strategic role that marine and coastal resources plays in Cuba's economic development must be clearly brought out so that national and local leaders and policy makers can make informed decisions with the right type of information. This has been the focus of our efforts. Several generations of Cuban marine researchers have worked on our ships and in our laboratories; nearly 30 million visitors have received information on general and specific marine issues; and our programmes on environmental education are included in the national educational system, from primary to university schools. Besides these outreach activities, we have instituted a system of national meetings (the marine environmental 'Olympics') for children, teenagers and elderly people that creates an intense movement of more than 50 thousand participants every year, looking for answers and solutions to their communities' main marine and coastal problems. The mass media are also important factors in the Aquarium programmes for capacity development. TV presentations, journals, thematic reports and debates of main issues and problems could not be successfully implemented without considering the important and necessary contribution of the media. Using mass media is our strategy to reach millions of Cubans.

'Learning by doing' and 'doing by knowing' are the key elements. Knowledge, information, education and science are the basis of all our actions. This is, in our experience, the only way to contribute to a real environmental education and, as a consequence, allow everyone to participate in the development of our country's 'capacity' as far as marine and coastal issues are concerned. The National Aquarium of Cuba has been fundamental in contributing to this marine and coastal environment understanding.

Kenya Meteorological Department

John G. Mungai, Ali Mafimbo and Joseph Mukabana

The Kenya Meteorological Department (KMD) provides marine meteorological services for the western Indian Ocean and is in the process of upgrading the Port Office in Mombassa to play a greater role in offering Marine Services to ships plying the Indian Ocean. In addition the KMD is the National focal point on Tsunami Early Warning over the Western Indian Ocean (WIO) region. We are increasingly involved in providing reliable and timely data and services to support political and managerial decisions in the marine environment.

Strategic planning and priority setting by KMD management now involves identifying marine challenges and threats that are important nationally. Prior to the 2004 Indonesian Tsunami, the marine division was inadequately staffed and equipped – three staff, no oceanographic equipment – and operated on a low budget. After the tsunami, the importance of marine activities in the department was prioritized internally, and recognized by the government, resulting in greatly enhanced marine equipment, more skilled staff and enhanced marine programmes. This included the installation of three automatic tide gauges and we are also in the process of acquiring a fixed buoy to be deployed in the Western Indian Ocean. Policy makers also agreed to the hiring of new staff, an unprecedented step since the employment embargo of 1997.

The KMD is now primed to support decision-making in marine management, and is currently undertaking training programmes to fill gaps in the new technical and management areas. In this respect, the Department has been involved in the IOC capacity development and leadership programme and various areas of technical training. This includes training staff in the fields of tsunami early warning and in physical oceanography. Two KMD members of staff have just completed their Masters in Applied Marine Science from the University of Cape Town.

The change of status of the KMD from a government body to a semi-autonomous one, in line with global trends, is expected to greatly improve the KMD's ability to carry out cost recovery programmes to diversify its funding base and directly relate to user needs.

Internationally, the Department is actively involved in efforts to set up an ocean observing system in the Indian Ocean and will use data from this system in addressing the issue of tropical cyclones and the impacts they have on western Indian Ocean sea surface temperatures and consequently on the climate and human activities of the region.

Factors in the successful development of Indian oceanography

Kalidas Sawkar and Ehrlich Desa

Fifty years ago the international community identified a gap in our knowledge of the Indian Ocean and laid out a vision that went beyond a planned expedition.

Oceanography in India has flourished since the founding in Goa in 1965 of the National Institute of Oceanography (NIO), its core ocean sciences institute. Within 20 years of its founding, its director was appointed to lead the newly created Department of Oceanography. The NIO became the lead agency in many national initiatives – organizing the country's expeditions to Antarctica (1981 to 1985), and surveying and laying claim to a deep sea site for mining poly-metallic nodules (1985). The requirements of the UN Law of the Sea provided another opportunity for NIO and other national institutes to apply their expertise for national causes.

We examine a case history of the factors affecting the birth and growth of India's National Institute of Oceanography, tracing its origins back to the International Indian Ocean Expedition (IIOE, 1959 to 1965) and noting the outreach it has today.

Three important factors responsible for the establishment of a strong growth trajectory for oceanography in India were:

1. *The needs of the region and of the international community were championed by respected scientists*: Dr N K Pannikkar, founder director of Indian fisheries institutes, and Professor Eugene LaFond, visiting professor at Andhra University, India, were two respected scientists familiar with Indian and international needs. Organizers of the IIOE motivated regional and international participation by highlighting practical aspects both for fisheries and the monsoons. The planning of the IIOE, following what we today recognize as essential *principles of engagement* in capacity development, ensured the success of the post-IIOE phase in India.
2. *High political patronage in the host country ensured visibility and resources*: Eminent Indian scientists from the emerging areas of space and nuclear sciences championed the cause of oceanography. They were responsible for the high patronage accorded to science and technology, including oceanography, from successive Indian Prime Ministers.
3. *Commitment of national resources and diversification of the ocean remit*: What started as one nascent organization in India has multiplied to several institutes dealing with ocean technology, ocean-based hazards, ocean data and polar research as Indian interests in her coastal and regional seas increase and diversify.

This growth trajectory is best demonstrated by the many new oceanographic institutes and improved infrastructure being created in India (including research vessels, special oceanographic satellites and deep sea monitoring buoys). As the IOC and the NIO approach their 50-year anniversary, the vision that launched the International Indian Ocean Expedition comes full circle, and Indian institutes are now able to partner countries, from Myanmar to Trinidad and Tobago, to survey and catalogue their offshore waters. The seeds that were planted half a century ago are truly self-replicating.

5

The early days and evolution of the Intergovernmental Oceanographic Commission

DESMOND P D SCOTT AND GEOFFREY L HOLLAND

As a young UK naval officer, Desmond Scott was involved in the early days of the Intergovernmental Oceanographic Commission. Later, from 1970 to 1979, he was Secretary of the IOC, in Paris. Subsequently he was Permanent Secretary for GEBCO. From 1985 to 1989 he served as Head of the UK Delegation to the IOC.

Geoffrey Holland worked for over 30 years for the Canadian Government in national and international ocean science affairs. In addition to his national duties as Director-General Ocean Science and Services, he participated in many intergovernmental and international organizations and has been associated with the IOC since 1970.

Scientists have worked across national borders over many centuries in the study of global phenomena. Even as early as the 1830s Gauss and Weber set up a global network of magnetic observatories. Cooperation in marine science among governments was first formalized with the establishment of the International Council for the Exploration of the Sea (ICES) in 1902 focusing on fisheries science in the North Atlantic.

Ideas for an intergovernmental organization to foster and benefit from coordinated studies of the oceans on a global scale began to form in the 1950s, eventually leading to the creation of the Intergovernmental Oceanographic Commission. This UN organization was given functional autonomy; however, recognizing the small scale of the initial enterprise and the limited funding

Figure 5.1 Heavy rainfall over the Maldives. One of the aims of the International Indian Ocean Expedition was to study the link between the ocean and the monsoons. © Filip Fuxa/ Shutterstock.com.

available, it was judged necessary to house it under the aegis of a larger UN agency, UNESCO.

The establishment and evolution of the IOC as the UN intergovernmental arm for marine science has had many setbacks, but there have also been notable successes, as recorded in many places in this book.

IOC history

For a more comprehensive account of the history of the IOC, see the document prepared to celebrate the 20th Anniversary of the Commission *A Focus for Ocean Research Intergovernmental Oceanographic Commission History, Functions, Achievements*, by Dr Hans Ulrich Roll. A more recent history of the IOC, by Geoff Holland, is in the UNESCO publication *60 Years of Science at UNESCO.*

As an intergovernmental body responsible for global ocean science and services, its resources have always been relatively meagre and as a consequence not always able to address effectively the needs of governments. Through all the developments, the general goodwill among national delegates, together with the indomitable determination and good humour of the Secretariat, has surmounted many difficulties. Innumerable lifelong friendships have resulted from these interactions over many years.

In this chapter we look back on some of the steps in the evolution of the first intergovernmental marine science organization in the UN, illustrating some human aspects through recollections of the people involved.

The beginning

The first documentation of ocean interest at UNESCO was at its 8th session in Montevideo, Uruguay, in 1950. The session authorized the Director-General to promote the coordination of research on a range of scientific problems including oceanography and marine biology. This initial impetus led to the formation of a small International Advisory Committee on Marine Science (IACOMS) composed of nine marine scientists appointed by the Director-General from a list of honorary consultants developed by the UNESCO Secretariat in consultation with FAO.

The function of this committee was to advise the Director-General on the promotion of international collaboration in the conduct of marine science programmes, taking into account the related work of other UN specialized agencies. In the following years IACOMS held several meetings and worked closely with another group of scientists established by the International Council of Scientific Unions (ICSU); this non-governmental body was the Special (later Scientific) Committee on Oceanic Research (SCOR). The partnership between UNESCO/IOC and SCOR has continued over the years and SCOR has just celebrated its own 50th anniversary. It was very clear that cooperative efforts were needed to tackle large-scale ocean science projects, especially in areas where little regional capacity existed and at its first meeting in 1957, SCOR initiated planning for an International Indian Ocean Expedition that UNESCO agreed to co-sponsor.

The Indian Ocean was considered an important but poorly understood region of the world oceans. The seasonal change of the monsoons and the interface between atmosphere and ocean presented a significant scientific challenge. Moreover, at that time, the surrounding countries were ill-equipped to tackle the logistics and research required. It soon became clear that an expedition of this magnitude would not be possible without national and intergovernmental commitments of ships, instrumentation, facilities and scientists.

Thereafter events moved relatively quickly. In November 1958, UNESCO made the decision to convene an Intergovernmental Conference on Oceanographic Research, which was held on 11–16 July, 1960. UNESCO did not act alone and many

Figure 5.2 Roger Revelle, one of the 'founding fathers' of the IOC. © IOC.

international and intergovernmental organizations assisted in the preparation of the Conference and participated in the proceedings. Many of these organizations would become important partners in the future programmes of the IOC. Among those present were representatives from the International Atomic Energy Agency (IAEA), Food and Agriculture Organization (FAO), World Meteorological Organization (WMO), World Health Organization (WHO), Intergovernmental Maritime Consultative Organization (IMCO), International Civil Aviation Organization (ICAO), International Council of Scientific Unions (ICSU), International Union of Geodesy and Geophysics (IUGG), International Association of Physical Oceanography (IAPO), Special Committee on Oceanic Research (SCOR), International Hydrographic Bureau (IHB) and the International Council for the Exploration of the Sea (ICES).

The principal recommendation to emerge from the discussions was that an Intergovernmental Oceanographic Commission (IOC) be established within the framework of UNESCO. This recommendation was taken to UNESCO and adopted later the same year by Resolution 2.31 at the General Conference, together with approval of an initial set of Statutes and an Office of Oceanography within UNESCO, to act as the IOC Secretariat. The justification for the birth of this new and valuable United Nations (UN) organization was based on the need for international cooperation in ocean research:

The oceans, covering some 70 per cent of the Earth's surface, exert a profound influence on mankind and even on all forms of life on Earth ... In order to properly interpret the full value of the oceans to mankind, they must be studied from many points of view. While pioneering research and new ideas usually come from individuals and small groups, many aspects of oceanic investigations present far too formidable a task to be undertaken by any one nation, or even a few nations. (UNESCO, 1960).

By this act, ocean science took a major step forward in terms of intergovernmental political visibility.

Despite the scientific origins of the IOC, it is also very likely that the governmental interest and support of the new organization was the result of a growing awareness of the strategic and economic importance of ocean science and information and its usefulness in the resolution of a broad range of national, regional and global issues.

Figure 5.3 Dr Anton Frederick Bruun of Denmark was the first Chairman of the IOC, but unfortunately did not live to complete his term of office. © IOC.

The control of the world's oceans had always been, and remains, central to confrontations over national security, trade and resources. The significance of ocean information and research had been demonstrated in surface and subsurface military operations during the war years and the presence of senior naval representatives on national delegations at these meetings was not only expected, it was also very necessary. The navies could make good use of information and knowledge arising from non-military operations, but the reverse was also true and, in fact, the capabilities of many countries in ocean science were based in military establishments.

In addition to the military interests, it became obvious that governments were looking to a new ocean organization as being more than a meeting place to discuss ocean research and to plan cooperative oceanographic experiments. The thrust of discussions at the early preparatory meetings and continued over the lifetime of the Commission has been to evolve beyond ocean research per se. It is not surprising therefore, that the IOC has continued to move towards the exploitation of ocean knowledge and information for the use and benefit of national governments and for collectively addressing national, regional and global problems, such as coastal management, ocean health, climate change, ocean services and capacity-building.

Just over a year later the IOC met for its first intergovernmental session in Paris at UNESCO Headquarters from 19 to 27 October 1961. Following the original proposal by Roger Revelle the IOC programme was divided into Ocean Science, Ocean Services and Training and Education. By the end of the session, a total of 40 states had become members of the Commission. This initial session of the IOC has a distinct place in its history and demonstrated the future usefulness of the Commission to organize regional and global ocean research programmes and to facilitate cooperation in important areas, such as developing standards and formats for ocean observations and data exchange, and debating the issues surrounding the deployment of fixed and floating ocean platforms. The discussions highlighted many concerns that would recur throughout the succeeding decades. Marine pollution, capacity-building and the availability and exchange of data were high on the list of initial priorities and have remained so. The discrepancy between the small size of the organization compared to the breadth and complexity of its intergovernmental responsibilities was evident in the early years and has not diminished to the present day.

Even at the first meeting, non-scientific politics diverted attention away from the agenda. The Soviet delegation expressed regret at the absence from the Commission of the lawful delegation of China, and questioned the right of the representative of Taiwan to represent China at the meeting. Political issues are an inevitable element of UN organizations and the meetings of the IOC were not protected by its mandate of ocean science. Nor were the politics reserved for governments alone; existing UN Agencies had jurisdictional concerns with the extent of the responsibilities of this newly formed ocean organization. In his welcoming speech at the inaugural session, the Acting Director-General of UNESCO, Mr. Rene Maheu, stressed that it was not the responsibility of the IOC to examine problems in meteorology, fisheries and other areas that came under the purview of existing UN Agencies, although to be fair, he did instruct the Commission 'to cooperate closely with other institutions of the United Nations family'.

Non-science issues for the IOC Secretary

Desmond Scott

Whilst the political aspects of the Law of the Sea were being studied and considered, and a legal regime for ocean space was being formulated by the United Nations, it was felt strongly that it was essential for the IOC to avoid political issues so far as this was possible. However during my term of office as Secretary of the IOC, there was one episode, during an IOC Assembly opening session, which can now be recorded without giving offence. It should be understood that Assembly seating arrangements were always in French alphabetical order. It was just the task of the Secretariat to inform the administration of the names of the countries and organizations sending delegations and the number of seats required for each delegation. Also it was at a time when feelings were running high against the apartheid regime in South Africa. Before one Assembly session, Professor Roll, then leader of the German (Allemagne) delegation, indicated that he was tired of sitting in the front row close under the podium and asked whether other arrangements could be made. This seemed a reasonable request and I changed the seating so that the letter 'B' (Belgique) started in the front row, with the idea of moving one letter down on each occasion. What had not occurred to me was that this also moved Afrique du Sud from the front row to the back of the room, and I was roundly accused by Algérie of trying to make the delegation of South Africa less conspicuous. The Assistant Director-General for Science was sent for and all the seating arrangements had to be changed back before the session could start. Professor Roll had to accept my failure to meet his request.

Reminiscences

Desmond Scott

IOC Chairman, George Humphrey, used to practise yogic meditation on the floor of his office during the lunch break. One day he was rudely brought back to reality by the screams of a secretary who had walked in and thought he was dead.

I remember the IOC circular letter, which had the words 'hesitate to' missing in the final sentence which therefore read: 'If you require further information, please do not contact us.'

I received a complaint from one of the female staff that the ADG for Science had instructed her not to wear a trouser suit in the office. There was a strong rumour that a UN Administrative Circular had been issued to this effect. How times change!

Many of the personalities that would influence later years were present. The first secretary was Warren Wooster (USA), who had also played a large role in the Copenhagen Conference; also in attendance were Konstantin Fedorov (USSR), Desmond Scott (UK) and Mario Ruivo (Portugal), three scientists destined to become future Secretaries of the Commission. In fact, Mario Ruivo has been a constant force in the work of the IOC and was very recently one of its Vice Chairs. Two future IOC Chairs, George Humphrey (Australia) and Nedumangattu Panikkar (India), were also participants, along with many of the leading ocean researchers of the time, including Roger Revelle (USA), Henri Lacombe and Jacques Cousteau (France), and George Deacon (UK).

The adoption of the IOC by UNESCO allowed it to take advantage of the existing meeting facilities at UNESCO Headquarters and to use the administration and framework of an established UN organization. UNESCO also endorsed the concept of functional autonomy for the Commission, which was seen to be essential to the advancement of its work and programme.

The evolution of responsibilities

The IOC enjoyed a relatively high level of recognition within the UN System during the first decade of its existence. In December 1966, the UN General Assembly

Figure 5.4 Despite the importance of ocean information to military intelligence, scientific cooperation continued in the intergovernmental arena throughout the cold war. © ppl/Shutterstock.com.

passed a resolution (2172 XXI) requesting the Secretary-General to make proposals to ensure the most effective arrangements for an expanded programme of international cooperation in terms of understanding the oceans and developing its resources. These proposals led to the development of a Long-term and Expanded Programme of Oceanic Exploration and Research (LEPOR) produced in cooperation with the FAO and UNESCO/IOC and endorsed by the UN (2414 XXIII). For many years LEPOR formed the basis for the scientific activities within the IOC.

It is interesting to note that the IOC, when dealing with this UN directive at its fifth session in 1967, considered a report 'International Ocean Affairs' that had been prepared by its advisory bodies. The report recommended that the Member States of the UN and its relevant Agencies give consideration to the establishment of a central intergovernmental oceanic organization to deal with all aspects of ocean investigations and the uses of the sea. At this time the Member States did not have the resolve or resources to accept such a challenge and concluded that a major change to the existing organizational arrangements was premature.

The UN Resolution endorsing LEPOR also urged Member States at the UN and relevant UN Agencies to agree, as a matter of urgency, to broaden the base of the IOC so as to enable it to formulate and coordinate such an expanded programme. Another important step in the development of intergovernmental coordination in ocean sciences was taken at the same session with Resolution 2467 XXIII that welcomed the concept of an International Decade of Ocean Exploration (IDOE) and requested the IOC to coordinate this activity in cooperation with other organizations.

In 1969, the IOC requested the Director-General of UNESCO to negotiate a formal basis of cooperation with other UN specialized agencies with interest in matters related to ocean science. The result of the negotiations was the establishment of a

The trials and tribulations

Ray Griffiths

An international civil servant sometimes has to work under really difficult conditions. When a meeting is held in a Member State, a formal agreement or letter of understanding is drawn up by UNESCO, on behalf of the IOC covering all the physical arrangements.

... a regional meeting of experts on marine pollution in Penang 1976, Malaysia, at that time, there were no computers, no high-speed photocopying equipment, and white printing paper was almost a luxury. Documents to be produced in situ, especially the Draft Summary Report of the meeting, required typing on waxed paper and run through a copying machine. ... minor editorial mistakes could be taken care of by painting a sort of nail varnish (it looked like and smelled like nail varnish) over the offending error and retyping the correction ....

... in Abidjan, 1978, we had a meeting room, facilities and supporting staff. I also had my secretary, Wendy Kerwin, with me, but no room or typewriter had been assigned to her. My Ivoirian counterpart – a gentleman by any standard – reassured us his man could handle all the typing. My secretary quietly told me that she had discovered the person assigned to type our texts. He was an elderly gentleman, full of good will, but with poor English and terribly slow. ... the solution! The American Embassy had a storeroom full of typewriters. 'Take your pick', they said. We did. Wendy worked in her hotel room each night, so as not to display our necessary deceit. We managed translation between us with the welcome but unrequested help of one of the interpreters ....

unique committee, called the Inter-secretariat Committee on Scientific Programmes Related to Oceanography. It consisted of the Executive Heads of the UN, the Food and Agriculture Organization (FAO), the World Meteorological Organization (WMO) and the Intergovernmental Maritime Consultative Organization (now the International Maritime Organization); the UN Environment Programme (UNEP) joined in 1972. ICSPRO was chaired by the Director General of UNESCO. A staff member was seconded from each of the FAO, WMO and IMCO to facilitate the cooperation and in 1974, approximately one-quarter of the IOC staff salaries and operational funds were provided by the ICSPRO Agencies. Unfortunately, by the

Figure 5.5 Since its establishment in 1960, the Intergovernmental Oceanographic Commission has been hosted by UNESCO in Paris and its membership has grown to over 130 Member States. © UNESCO.

mid-1970s, the financial constraints throughout the UN system became more apparent and the ICSPRO arrangement faltered; nevertheless it was an interesting attempt at interagency cooperation.

Searching for a solution

It has not been an easy task for a small intergovernmental body within UNESCO to exercise a global mandate in ocean sciences and services. Obviously, the organization can neither operate scientific programmes, nor provide financial aid to assist developing countries in achieving ocean science capacity, nor can it act as an operating agency in the provision of ocean services. The programmes generated and coordinated within the IOC are implemented through the resources of the Member States themselves. The IOC Secretariat undertakes to organize meetings of experts, programme coordination, working committees, training sessions, etc. and of course the decision-making meetings of the Assembly and Executive Council. Over the years the Member States have struggled with ways of coping with the expanding responsibilities of the Commission and the constraints imposed by the

Figure 5.6 Moving into the electronic age had its problems. Image courtesy of Gary Bugden, Bedford Institute of Oceanography.

lack of financial and staff resources. Regional considerations have added to the burden. Member States from different ocean regions have their own priorities and programmes to be pursued and it makes sense for these to be organized on a regional basis with the countries concerned. The two most prominent Regional Commissions of the IOC are the WESTPAC and IOCARIBE representing the Western Pacific and Caribbean regions respectively. Nevertheless, servicing regional secretariats imposes an additional burden on the IOC structure that has never been satisfactorily resolved.

Over the years Member States have studied the operation and function of the Commission, often working in *ad hoc* intergovernmental groups meeting intersessionally. Most notably have been the report 'Quo Vadis IOC', the recommendations of the 'Development, Operations, Structure and Statutes' (DOSS I and II) and recently the study, 'Future of the IOC'. The results have been debated at the governing bodies and have resulted in changes to the Statutes and workings of the Commission. As the mandate and responsibilities of the IOC continue to expand and change, it is to be expected that such internal examinations will continue.

The demise of ICSPRO

Desmond Scott

There was one occasion when all the (representative) members were attending other meetings in FAO, Rome, so I took the opportunity of convening a meeting of ICSPRO at the same location, well in advance with all the necessary documentation. Then at the appropriate time I took my seat as Secretary of the Committee. A few minutes later a head appeared round the door: 'Could I possibly look out for the United Nations', this was shortly followed by another head, 'Could I possibly look out for FAO', and so on until I was left on my own representing the UN and all the relevant agencies. It is said that a one man committee is the best way of achieving consensus and useful action, but this was hardly the intention of the IOC founding fathers when ICSPRO was set up.

The maturing process

The many self-searching exercises undertaken by the IOC noted above have resulted in changes to its mandate, duly recommended by the Member States and endorsed by the UNESCO General Conference. Drafted in 1960, the Statutes of the Commission were revised in 1970, 1987 and 1999 as changes were found to be necessary to reflect its broadening function and responsibilities.

The latest proposed amendments to these Statutes were adopted by the IOC Assembly at its twentieth session and following approval by UNESCO came into force on 16 November 1999. The present Statutes demonstrate the growth of the Commission from its initial establishment as a cooperative mechanism for the conduct of large-scale ocean research programmes, to a fully fledged UN Specialized Agency with programmes and responsibilities affecting the programmes and policies of governments on a continuing basis.

The purpose of the Commission now reads '... *to promote international cooperation and to coordinate programmes in research, services and capacity-building, in order to learn more about the nature and resources of the ocean and coastal areas and to apply that knowledge for the improvement of management, sustainable development, the protection of the marine environment, and the decision-making processes of its Member States.*'

Figure 5.7 Ocean science is needed whether the problem is controlling pollution or planning for offshore renewable energy. © tlorna/Shutterstock.com.

Figure 5.8 Intergovernmental ocean science may be concentrated in relatively few countries, but countries at all levels of development need to be able to protect and sustain their marine resources. © urosr/Shutterstock.com.

Figure 5.9 The huge offshore oil and gas industry growth was not a factor in the establishment of the Intergovernmental Oceanographic Commission 50 years ago. © Yvan/Shutterstock.com.

In addition the Commission '*will collaborate with international organizations concerned with the work of the Commission, and especially with those organizations of the United Nations system which are willing and prepared to contribute to the purpose and functions of the Commission and/or to seek advice and cooperation in the field of ocean and coastal area scientific research, related services and capacity-building.*'

The latest formulation of the Statutes also holds an interesting addition relating to the continuing problem with the lack of resources. Paragraph 4 of Article 10 (Financial and other resources) reads '*The Commission may establish, promote or coordinate, as appropriate, additional financial arrangements to ensure the implementation of an effective and continuing programme at global and/or regional levels*'. This paragraph gives the IOC the responsibility to pursue alternative means of obtaining resources for its programmes. The implications of such authority have not yet been fully explored.

Over the years the Commission has not neglected the need to maintain an up-to-date scientific basis for its programmes. In 1975 the Assembly agreed to establish a Scientific Review Board (SAB), which met in the following years and recommended a study be undertaken by the IOC, in collaboration with its advisory bodies, on ocean science to the year 2000. An expert consultation was held in Villefranche-sur-Mer, France, 13–17 April 1982, with leading oceanographers of the time, and a report prepared and submitted to the twelfth Assembly the same year.

A similar exercise was organized 20 years later in collaboration with SCOR, and the Scientific Committee on Problems of the Environment (SCOPE). This latest assessment was carried out by scientists nominated by their peers and selected by the three sponsoring organizations. The experts met in Potsdam, Germany, in October, 1999 and the resulting publication 'Oceans 2020' was released in 2002.

Taking the long view

David Pugh

Three distinguished some-time Chairmen and an Executive Secretary were talking to the very attractive lady delegate from a central American country at a Russian reception. In the politically incorrect way of those times, they complimented her on the way she glamorized the governing body proceedings. They assured her that whenever she entered the room they all attempted to smarten themselves up: ties straightened, hair smoothed back, collars brushed … to which she replied 'I don't know why you bother, I can't see a thing until I put my glasses on'.

It's all about communication

Geoff Holland

There is an enormous contribution to intergovernmental cooperation from the dedication of the army of interpreters and translators dealing with six different languages. We don't always make it easy for them. Coming from a bilingual country and having acquired my French at a late stage in life, I nevertheless took the plunge and made two interventions in my non-mother tongue at an IOC Assembly. My self-esteem was shattered, however, when I was approached by a long-suffering interpreter in the coffee break who whispered a word of advice in my ear… 'Please … stick to English!'

The changes to the operation and function of the IOC have not arisen solely from the requirements and priorities of its Member States. Many external forces would prove to have an influence upon the ocean community in general and the IOC in particular. The political recognition of the importance of the environment and its place alongside the economy and health in dealing with human development was one of the first to emerge. In 1972, the UN Conference on the Human Environment was held in Stockholm to draw attention to the planetary environment and the global issues that needed to be addressed by society. The oceans were not a large part

Figure 5.10 Prince Albert I of Monaco and crew posing on the deck of the *Princess Alice* with a dissected cetacean in 1912. Prince Albert was a pioneering marine scientist. © NOAA.

of the agenda, but several recommendations focused on ocean pollution, with a specific request to the IOC to create a programme for the investigation of pollution in the marine environment, a request that was already under consideration by the IOC as part of LEPOR.

Twenty years later the Stockholm initiative was followed by the Rio UN Conference on the Environment and Development (UNCED), an historic meeting and one that would influence greatly the evolution of environmental programmes over the succeeding years. UNCED gave rise to an environmental agenda, Agenda 21, which included a Chapter 17 specifically dealing with Oceans. The Agenda demanded that an integrated and comprehensive global ocean observing and information system be created to contribute to oceanic and atmospheric forecasting, ocean and coastal zone management and global environmental change research. There were recommendations related to capacity building, coastal protection and management and others that called for an examination of the effects of global changes on the ocean. The influence of Agenda 21 on the IOC and on other UN organizations concerned with the environment was substantial. Again, and perhaps understandably, the IOC had already taken steps to address many of these issues.

In 1995, governments agreed on a global plan of action, backed up by national plans, to address the protection of the marine environment from land-based sources of pollution, mostly the impact caused by the run-off of agricultural, industrial and human wastes into the vulnerable coastal waters.

The 2002 World Summit on Sustainable Development (WSSD) took place in Johannesburg to review the progress in the implementation of Agenda 21. The IOC undertook a proactive approach to this latest world conference on the environment. The Commission approved a message from its Member States to the Summit and submitted a document entitled 'One Planet, One Ocean' describing the Commission,

Living with diversity

Ray Griffiths

… in Libreville, 1979, we had been promised a meeting room for about 20 regional experts. When we arrived we were informed that the government had simply asked the Senate to meet elsewhere, while the IOC meeting was in progress. The Senate in fact occupied a room designed to welcome a meeting of the Organization of African Unity. We were flattered, but the room was in the form of a circle of 30 or more metres in diameter and our twenty participants were spread around the 100-metre circumference! If the Libreville meeting room was a surprise, the one proposed for the regional marine pollution meeting in Montevideo in 1980 was a shock. We were asked to conduct the meeting at a table about the size of a full-size billiard table on the stage of the Teatro Nacional. I protested strongly. How could they have agreed to such a set-up for a meeting of regional experts? Well! The Navy was greatly interested in promoting public awareness of marine pollution problems and it would create a great disappointment locally if the meeting was not open to the general public. So we 'performed' our meeting to an audience of two hundred or more members only too happy to satisfy their curiosity, free of charge.

… 1988 regional meeting in Lahore, in the foothills of the Himalayas … One participant – a redoubtable lady scientist of Pakistan – requested the Chairman – a compatriot of hers – to declare that the meeting be held under Shari'a Law … no problem. Under the new and unexpected legal regime, a mullah, each morning before the day's business got under way, read a verse or two from the Koran, in Arabic, as is the custom. I later approached him and asked him if he could find some verses that dealt with the ocean and he said he would see what he could do, but I will never know – my command of Arabic approaches the reciprocal of infinity.

together with other documentation explaining how its programmes related to the goals of sustainable development.

Another huge change in ocean affairs that has had significant impact on the programmes and policies of the Commission, over the last part of the twentieth century, was the emergence of the 'new ocean regime'. This change began in earnest with the third UN Conference on the Law of the Sea, held in Caracas in 1974, and

Keeping a proper perspective

David Pugh

A sessional drafting group was preparing a Statement for the 2002 Johannesburg Summit. As Chair, I was being bombarded with suggestions for amendments. In desperation I protested that I could hold only five things in my head at one time! From the end of the table, in his usual impeccable English, the French delegate said, very quietly 'Napoleon could manage only four'.

culminated with the Convention coming into force in1994. The IOC is recognized within the Law of the Sea Articles as the 'competent international body' dealing with ocean science. This recognition gave the Commission a substance and prestige in ocean law that would be cited time and time again as a proof of its stature and responsibilities. Unfortunately, the interpretation of what could or should be expected of the IOC in the light of these explicit and implicit responsibilities has proven to be a difficult task and the full implications are still being assessed.

The context within which the IOC was established has changed and the Commission has adapted accordingly. As the needs of national governments, regional arrangements and global cooperation in ocean issues evolve in the future, the Commission must continue to monitor its ability to respond and, as necessary, further adapt its mandate as the UN organization responsible for ocean science.

Part III Oceans and Science: Preface

In terms of the theme for this book it should be realized that governments support ocean science both directly and indirectly. In its widest definition it encompasses the whole spectrum from pure to applied science and is undertaken in both government laboratories and in academia. Ultimately the public purse is responsible for both although the priorities of the former will be dictated more strictly by attention to economics and social services and the latter to addressing education and the extension of boundaries to research. Eventually, both comprise the field of ocean science, which provides the conduit between the supply of data and the information delivered to marine managers and decision-makers. The important role of non-governmental international ocean science must also be mentioned. To a considerable extent such organizations provide an independent and peer review of issues. They may be driven by purely academic approaches to science, as in the case of the Scientific Committee on Oceanic Research, by environmentally generated concerns, as in the case of Greenpeace, or by bringing together a mixture of governments, non-government organizations, United Nations agencies, companies and local communities as in the case of the International Union for the Conservation of Nature to develop and implement policy, laws and best practice.

Ocean science can be driven by the demands of external priorities of social, political or commercial interests. Alternatively, science can generate developments and policies from within its own ambit, defining gaps in current knowledge that need to be addressed and frontiers of research to be mined and explored. Discoveries generate activities leading to the establishment of new observing systems, improvements in technology, methodologies and modelling techniques. An essential component is the spatial definition and corresponding physical and other metadata that allows the various and disparate data sets to come together seamlessly and usefully in the service of marine communities. In turn increased and more accurate information will generate feedback into management decisions and policies creating a dynamic interaction between scientific knowledge and its social and economic consequences.

Ocean science is an important element in intergovernmental cooperation for capacity building activities, providing a necessary tool for the development of marine economies and societal improvements in disadvantaged nations.

Ocean science is obviously a very broad area that cannot be covered comprehensively in our present text and of necessity the following chapters have therefore been selected as representative of the broader whole. The authors are experienced and expert scientists in ocean research and international and intergovernmental cooperation. An international organization, the importance of mapping, the evolution of ocean science in general and in the pursuit of solutions to ocean climate prediction and mitigation have been included alongside a very specific and more recent example of a successful ocean science application to the issue of harmful algal blooms.

Of the many regional and specific scientific organizations, one prominent marine body is the Scientific Committee for Ocean Research of ICSU, which recently celebrated the half-century since its founding in 1957. The same vision and the same people who set up SCOR, including notably Roger Revelle, from the Scripps Institute of Oceanography, were also the drivers for setting up the IOC as an intergovernmental organization for marine science a few years later. Cooperation between these two bodies over the intervening half-century have been mutually beneficial, with the occasional ebb and flow expected of common but somewhat differently defined goals.

The importance of the advice and direction given to governments and the intergovernmental community by independent international agencies cannot be understated. Whether these represent public opinion on environmental matters or the collective advice of academia, their influence on governments and governmental policy is substantial. In this book addressing ocean science, the authors have asked SCOR to write a representative contribution to recognize this critical non-governmental scientific element.

6 Ocean science: an overview

GUNNAR KULLENBERG

Gunnar Kullenberg has researched ocean mixing, air–sea interaction, ocean optics, colour and remote sensing, pollution, interactions in physical–biological processes, ocean circulation, coastal processes and upwelling, and coastal and ocean governance. He is a former Executive Secretary of the Intergovernmental Oceanographic Commission, and subsequently Executive Director of the International Ocean Institute.

The Context

This overview aims at presenting ocean science programmes in the context of trends, concerns, challenges and changes that have emerged during the past several decades. I am grateful to past IOC Chairman Ulf Lie for his assistance with this text.

The development of marine technology during, and directly following, World War II provided the means of discovering vast resources in the ocean and the sea floor. The ocean, until that time regarded essentially as a fishing ground and a means of transportation, became a supplier of very significant resources of potentially enormous importance for industry and economic development and for marine food production. Governments, recognizing these benefits, proclaimed extended national jurisdictions over shelf sea areas, which were pushed further and further seawards, to cover preferential treatment, exclusive rights, extended fisheries zones and even full sovereignty. Neither the Geneva Conventions of 1958 on the Law of the Sea nor the supplementary second effort in 1960 could stop the process. Nevertheless, the importance of the ocean and its resources for humankind was recognized. In association came the realization that marine science and technology could provide the foundation for strong economic development and that international cooperation was required to tackle the scientific challenges. Several, essentially national, ocean research expeditions materialized in the first decade after World War II, including the Swedish Albatross and the Danish Galathea. These stimulated the global international

Figure 6.1 Research ships have measured the sea for hundreds of years. Here a Chilean survey ship anchors off Easter Island (Rapa Nui). Image by David Pugh.

base for marine research through several new developments and discoveries. The cooperative study of the Planet Earth got a large boost through the success of the International Geophysical Year 1957–58. The process triggered the creation of SCOR, the Special (later Scientific) Committee for Ocean Research, by ICSU, which was followed in 1960 by the establishment of the intergovernmental IOC within UNESCO.

The forcing

The IOC, despite its modest size, represents the most important intergovernmental body in ocean science. It is driven by internal forcing from its Member States, and external forcing from interests, overall global developments and concerns outside its control. The IOC mandate has accordingly evolved with changing needs, with improved scientific and technological capability and concurrently influenced by national policy and international law developments. Our knowledge and understanding about the planet has grown gradually over the centuries, but at an increasing rate over the last 50 years. Discoveries continue to be coupled to needs of society related to resources, food, protection, transportation and military considerations. The Millennium Declaration of September 2000 states: 'We must spare no effort to free all of humanity, and above all our children, from the threat of living on a planet irredeemably spoilt by human activities, and whose resources would no longer be sufficient for their needs', reflecting a growing concern about our entire life-support system. The ocean is a key component of that system, through its impact on climate, the hydrological carbon, nitrate and phosphate cycles, biological diversity and the ecosystem. In addition, the ocean is a source of water vapour and freshwater, food, energy from many sources, non-living

Figure 6.2 A spectacular setting for polar survey work in difficult conditions. © IHO.

resources, genetics and medicine and underpins 80% or more of our trade. Therefore, knowledge of the marine environment is of the utmost importance for the future of humankind. The IOC was established to improve the use of human and infrastructure resources for this purpose using intergovernmental cooperation.

The ocean is a global commons and all nations are involved, not only through marine activities, but also activities on land are directly coupled to the land–ocean interface and thus the coastal and shelf seas and beyond. National jurisdiction over coastal ocean areas has gradually increased until today the ocean areas under national jurisdiction approach nearly half of the ocean surface. These new political boundaries have a profound influence on marine scientific research. Under the Law of the Sea articles, prior consent from the coastal State is required in order for international cooperative ocean research to be carried out in the Extended Economic Zone, thus departing from the freedom of scientific research paradigm and demanding the attention of all governments.

The UNCLOS provides a strong international instrument to help achieve equity and benefit sharing. It was presented by the Secretary-General of the United Nations as 'one of the greatest achievements of this century'. It establishes a regime by which

Coastal States can benefit from the resources of their continental shelf and provides a foundation for utilizing the ocean areas beyond national jurisdiction for the benefit of all. A future, however, that imposes stewardship obligations and requires knowledge and wise management based on scientific findings.

Other external forcing events are coupled to societal concerns about the degradation of the human environment, socio-economic conditions or global planetary changes, all of which generate challenges to science. These global issues were first reflected in the UN Conference on the Human Environment, Stockholm 1972, which was followed by the release of the Brundtland Report on Sustainable Development, in 1987, and the UN Conference on Environment and Development in Rio de Janeiro 1992. These events led to an agreed programme of action (Agenda 21) and Conventions on Climate Change and Biological Diversity, and several other instruments/agreements focusing attention to specific concerns or areas. The new millennium has seen the World Summit on Sustainable Development 2002 in Johannesburg, and the associated Johannesburg Plan of Implementation.

Planetary research requires international cooperation and ocean research is no exception. The trend since World War II has followed gradual centralization into large programmes, focusing on identified specific scientific problems, as well as pooling resources for new large infrastructure investments. This leaves fewer opportunities for individual scientific achievement, but provides for larger cooperative programmes and the support of centralized, high-quality infrastructures and facilities. In this way, very expensive and sophisticated experiments can be carried out in the ocean interior and on the seabed. Research on deep-sea vents, genetic resources, continental margin, the lithosphere, global observing programmes and basin-wide ocean–atmosphere and climate-related studies are examples. Intergovernmental cooperation is essential to ensure coordination, implementation and benefit sharing and is achieved through policy-oriented discussions on exchanges, participation, access, benefit sharing and the establishment of science advisory groups.

The first approach

In the early years and following the basic reasons for its creation, the IOC, in cooperation with international scientific advisory bodies and intergovernmental organizations, focused on stimulating development, coordination and implementation of several international cooperative regional studies. These studies needed an intergovernmental platform to attract the participation of the regional countries and to coordinate the resources of other interested countries. In turn these efforts stimulated other essential tasks of the IOC, such as assisting developing

Figure 6.3 Until the 1970s the primary method of collecting ocean temperature, salinity and chemical properties was by lowering instruments on a hydrographic line. Here a Knudsen bottle is being deployed. © NOAA.

countries to achieve capacity in marine science, encouraging ocean data exchange and establishing permanent international observation systems. Cooperation on regional process-oriented studies demonstrated cost–benefit returns to resource managers and government officials and led to an enhanced interest for marine research in the countries of the regions concerned. Training and educational considerations were important elements of these regional endeavours. Important advances in our knowledge of the oceans over this period were documented in the first assessment of progress in marine science, prepared by the IOC and SCOR in 1969.

The decade of the 1970s

The next decade continued the successful growth of ocean science with broader policy considerations. In 1968, the UN General Assembly adopted an International Decade of Ocean Exploration (IDOE, 1971–1980). In its leading role for the Decade the IOC gave more freedom to individual scientists and groups of the scientific community to generate programme elements while continuing to facilitate coordination and cooperation. The Decade was used to stimulate actions and raise support at national levels to participate in the international activities and to develop national programmes. It was a great success for ocean research, building a scientific foundation for the subsequent system-oriented and interdisciplinary research programmes that would follow.

Other external forces were also playing a role in enhancing global interest in the oceans. The speech of Arvid Pardo of Malta, to the General Assembly in 1967, triggered negotiations for the new Law of the Sea and increased governmental attention on ocean affairs. The Stockholm Declaration stressed the need for science and technology in relation to the identification, avoidance and control of environmental risks and highlighted the concern for marine pollution and the use of the ocean as a recipient of wastes. Responding to the environmental awareness, the Global Investigation of Pollution in the Marine Environment (GIPME) was formally accepted by the IOC as a major component of the IDOE and the GIPME Comprehensive Plan provided a scientific framework for marine pollution studies during the Decade and beyond. For many years the GIPME efforts were led and driven by the dedicated work of Neil Andersen of the USA. The marine environmental activities of GIPME engendered cooperation with other intergovernmental bodies interested in the use of the scientific results to address marine pollution issues and a Joint Group of Experts GESAMP was co-sponsored as an independent advisory group by many of the marine-related UN organizations.

The final two decades of the Second Millennium

While the scientific results laid the foundation for the further advancement of marine science, the institutional developments, those emerging from the Law of the Sea negotiations and the Brundtland Report were to significantly influence the marine research in the following decades. In parallel, concerns over climatic planetary changes related to human actions, in particular the population explosion, were gaining strength within the science community. This was evidenced through the creation of the International Geosphere Biosphere Programme, IGBP, of ICSU; the creation of the World Climate Research Programme (WCRP), in 1978/79, sponsored by WMO and ICSU; and the inter-disciplinary SCOR/IOC Committee on Climatic Change and the Oceans (CCCO) in 1978, the latter under the leadership of Roger Revelle.

The importance of interactions amongst the various components of the planetary environment, and the large role of the ocean in the climate system, was becoming evident. In turn, this emphasized the need to strengthen the role of the marine science community in the research programmes related to climate and global change problems. The IOC enhanced its association with the World Climate Research Programme through becoming a co-sponsor in 1991. The development of ocean climate research is extremely important and is fully developed in a separate chapter in this volume. In concert with the recognition of the importance of ocean science to the global environmental issues was the development of the necessary tools to collect the needed data for research and monitoring purposes. Traditional methods of using research vessels, though continuing to be necessary, were inadequate in terms of spatial coverage. Advances in automated and satellite observations yielded the necessary tools to overcome the lack of data that was constraining the understanding of ocean processes.

Figure 6.4 The oceans were once considered an ideal place to safely dispose of unwanted waste. Mustard gas canisters are shown being dumped into the Atlantic in 1964. Image from Rochelle F.H. Bohaty (2009) Lying in wait. *Government & Policy* **87** (13): 24-26. © US Army.

Not all the external influences on marine science were being driven by global considerations. The increased jurisdiction of Coastal States over the coastal waters and their resources was generating increased awareness of the importance of processes and circulation on the continental shelf and the need for improved understanding and management. The need for IOC to pay attention to and help address associated legal issues was recognized, and now fully acknowledged in its revised Statutes. Integrated coastal area management and the related science programmes became a priority for many governments and their concern was reflected in the development of related international science efforts, such as the Coastal Ocean Advanced Science and Technology Study (COASTS) and the Land-Ocean-Interaction in the Coastal Zone (LOICZ) programme. The interaction between marine law and ocean science was also demonstrated by the negotiations over coastal and continental shelf seabed resources, where the resolution of jurisdictional limits required a comprehensive knowledge of the underlying geology and again demonstrated the lack of capacity in many of the less developed Coastal States.

The environmental policies started in Stockholm were coalescing into more defined and comprehensive approaches to global issues and the underlying science. Scientists were paying more attention to interdisciplinary research on the one hand and governments were recognizing the value of 'ecosystem management' and 'one earth'. All these environmental and legal developments in the second part of the 1980s and early part of the 1990s fitted well the process of UNCED and the series of intergovernmental follow-up actions. Both the scientific community and governments realized that global change provided a framework within which local, regional and global scientific cooperative programmes should be viewed. In the marine area, this included paying attention to areas particularly vulnerable to hazards such as sea level rise, storm surges, tsunami, as well as continental shelf dynamics and harmful algal blooms. The Rio Conference highlighted the increasing public concern with respect to environmental and developmental problems. The need for proper coordination among the large number of activities to avoid duplication and ensure exchange and best use of limited resources was evident. Communicating the science results and linking the science efforts to the various intergovernmental negotiations and assessments was seen as a necessity.

Figure 6.5 Understanding the relationship between the ocean and the atmosphere is a critical element of weather forecasting. © Fedorov Oleksiy/Shutterstock.com.

The significance for ocean science of the entering into force of the Law of the Sea cannot be overstated. It accepts the importance of marine scientific research by addressing it at the same level as other major provisions of the Convention in a dedicated part XIII, stipulating the international rules for the research regime, including provisions regarding consent and role of international organizations. The IOC is given special responsibilities with respect to the Commission on the Limits of the Continental Shelf and on Special Arbitration and these responsibilities have been recognized in the revised IOC Statutes.

The importance of the role of the IOC to promote marine research, the understanding of ocean processes and the inclusion of ocean science concerns in international actions addressing sustainable development was highlighted in the International Year of the Ocean, adopted by the UN in 1998. The programme for the Year included several dedicated research cruises, national science conferences and scientific publications. Appropriately the preparation of the third assessment of ocean science was initiated in 1998 and completed with the publication in 2002 of

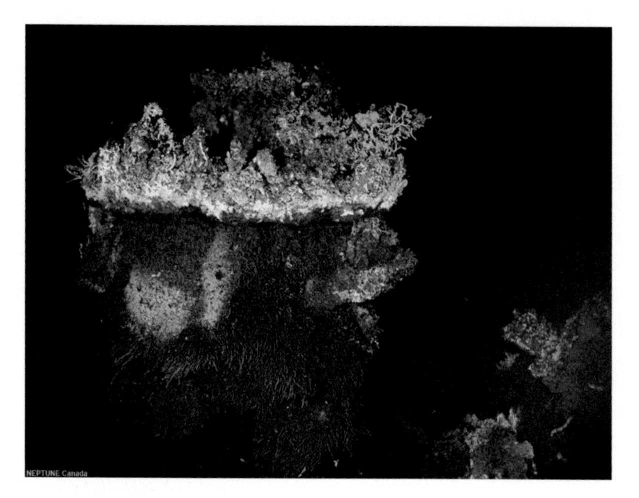

Figure 6.6 Life at great depth. A hot vent 'chimney' colonized by tube worms and wearing a sulphide crown precipitated from the superheated water. Image courtesy of NEPTUNE Canada.

Oceans 2020: Science, Trends and the Challenge of Sustainability, jointly with SCOR, SCOPE and IOC. The assessment acknowledged that 'recent progress in ocean science has been achieved through international coordination and planning'. It also confirmed the trend of ocean research moving towards establishment of sustained observation platforms at sea, paired with use of satellites for observations and data transmissions in real or near-real-time. The growing importance of powerful fast computer facilities for dynamic modelling of the physics of the ocean was noted, together with the associated emerging forecasting abilities over timescales of months.

The dawn of the new millennium

The new millennium has seen a crystallization of the programmes and policies that had been developed over the previous decades. The growing recognition of the importance of the oceans and marine science was evident at the UN with the creation of the UN Informal Consultative Process on Ocean Affairs and the Law of the Sea in 1999, a high-level mechanism reporting directly to the General Assembly. A Sub-committee for Oceans and Coastal Areas for the implementation of Chapter 17 of Agenda 21 facilitated co-sponsorship and joint implementation of many programmes, pooling resources and know-how from different sectors. The social, human concerns are manifested in the Millennium Declaration and the International Development Goals adopted in September 2000 by the UN General Assembly. They reflect the issues of human-social development, including: poverty reduction; education; reproductive health; environment; with a national strategy for sustainable development in the process of implementation in all countries, by 2015. The need to

Figure 6.7 The lives of many coastal communities around the world depend on a healthy sea and sustainable resources. © 'Bond Girl'/Shutterstock.com.

achieve these goals is of great significance for the ocean science agenda of the initial decades of the new millennium. They can only be achieved through cooperation, and they could act as a unifying theme. Hence the IOC functions as a joint specialized mechanism and its acceptance as an international organization for marine science appears cemented.

An important event early in the new millennium was the World Summit on Sustainable Development, in Johannesburg. The results of the science efforts following UNCED through the research related to climate, to coastal zone management and shelf seas conditions, ocean observations, and the UN Year of the Ocean were noted through the Declaration of the IOC, transmitted to the Secretariat, which stated that 'Sustainable development is highly dependent on the wise management of the oceans and coasts based on scientific knowledge'. A specific action forthcoming from the Conference was the call for a regular Global Assessment of the Ocean, which, if accepted, will have important consequences for future ocean research efforts.

The World Ocean Circulation Experiment (WOCE) analysis phase, completed in 2002, resulted in atlases and modelling advances on the basis of the vast amount of data and the process studies. These have been of great importance for subsequent activities relating to climate variability and predictability and global ocean data assimilation.

Marine science continues to face several basic questions: how large are the ocean carbon sinks and sources, how will they react to the increasing atmospheric content of carbon dioxide, and to the changing ocean conditions, how are they distributed and which processes control them and what are the affects on the ocean and its ecosystem? The urgency and significance of this research were noted by the G8 meeting in 2003, committing to improved coordination and cooperation among national and

Figure 6.8 Underwater observatories represent a frontier in ocean observation, able to manage a variety of remote *in situ* and mobile Instruments on-line. 'Wally' is an Internet-operated crawler managed by the Canadian NEPTUNE observatory. Image courtesy of NEPTUNE Canada.

international bodies including the IMO, FAO, IOC and UNEP, to strengthen science-based decision-making. This has led to the initiation of the International Ocean Carbon Coordination Pilot Project, coordinated by the IOC, co-sponsored by the SCOR, cooperating with the IGBP, and addressing the identified scientific uncertainties through research and a coordinated network of ocean carbon observations.

Our increased knowledge and understanding of the ocean and its processes is not only vital to the solution of environmental and climate related issues, but it also raises new concerns. For example, the ability of the ocean to absorb carbon dioxide is an essential feature of the climate considerations; however, the acidification of the ocean from this process, and its impact on marine ecosystems, is now also recognized as an issue. In addition commercial interests are interested in the economics of using the ocean for carbon dioxide sequestration, again with concerns from the scientific community about the impact of such activity on marine life. The IOC and SCOR have established and recognized the need to monitor the developments with respect to the carbon dioxide sequestration and established an Advisory Panel on Ocean carbon dioxide to watch developments. Several other climate studies are focusing on international ocean carbon research.

There is a growing emphasis on ecosystem research covering coral reefs, large marine ecosystems, harmful algal blooms and global ocean ecosystem dynamics. Ecosystem research highlighted the so-called regime shifts with observed physical and biological changes in the North Pacific. The basin and global scale modes of marine ecosystems brought about a gradual integration of research from several different programmes as GLOBEC, JGOFS and WOCE, helping to bring together observational and theoretical ecology. Coupled to this the marine pollution research saw a focusing on human activities potentially causing adverse effects in the marine environment and the ecosystems, through two core activities: transport, cycling, fate and effects of contaminants; and indicators of marine environmental conditions. The Joint Global Ocean Flux Study entered its synthesizing phase following 10 years of observations. This includes the contribution from continental margins and seas to carbon dioxide sequestration and the horizontal flux of carbon, phosphorus and nitrogen across the ocean–continental margin boundary.

An operational milestone for the ability to implement an integrated observing system was the creation of the IOC-WMO Joint Technical Commission for Oceanography and Marine Meteorology (JCOMM) having its first session in 2001. This marks not only the mutual acceptance by member states of the two organizations as equals, but also that of the respective science communities in realizing their mutual responsibilities in addressing many challenges of great social concern. One would wish that a similar acceptance of the need for cooperation in addressing the challenges could be generated between the salt- and fresh-water communities, hydrology and oceanography. It can be noted that it took several years to achieve the relationship allowing the establishment of the Joint Commission.

A large part of the continuing marine science requirements remain focused on local and regional issues. National priorities contain adaptation and mitigation requirements related to climate change, environmental issues and the management of conflicting marine activities. The interaction between climate and integrated coastal area management research has resulted in a regional project 'Adaptation to climate change – responding to shoreline changes and its human dimensions in West Africa through integrated coastal area management'. Related to this is the increasing attention to research on ocean ecosystems and environmental protection, evidenced through new focus areas and combinations of marine pollution, living and non-living marine resources, including studies of algal blooms; coastal benthic systems; coral reef bleaching; global ocean ecosystems dynamics; land–ocean–atmosphere biogeo-chemistry, involving nitrogen, phosphorus, silica fluxes and their modelling, together with coupling between fish recruitment and environmental conditions.

Figure 6.9 A hydrate glacier on the ocean floor. Most of the natural gas-containing hydrates are in the ocean bottom, and it is speculated that eventually this fuel source will be exploited to meet growing energy demands. Image courtesy of Ross Chapman.

Understanding the relationship between biology and meteorological and oceanographic parameters will provide a basis for monitoring, forecasting and early warning of changes. Research provides further insights into the services yielded by the ecosystems, in carbon storage, coastal protection, provision of medicine and genetic resources, stressing also the need for research efforts on marine biodiversity. The latter triggered the Census of Marine Life Programme, initially a public endeavour and recently adopted by the IOC as a cooperative governmental programme. The increasing knowledge about and concern for coastal hazards from the ocean, placed in focus by the tsunami of December 2004, has led to digestion of the scientific information on storm surges, flooding, tsunami-related research as well as that on the longer term sea level change, resulting in guidelines for coping with such hazards.

Addressing challenges: strategic adjustments and forging coalitions

An IOC Advisory Group on ocean science suggested focusing on impacts of climate variability and change in the marine environment and on its living resources and ecosystems. This includes explicit recognition of coastal research as a primary element, including climate impacts; direct anthropogenic influences, integrated coastal area management, natural marine hazards and coastal protection; marine assessments with emphasis on the science that will underpin the proposed Global Regular Assessment of the Marine Environment; and inclusion of a cross-cutting element in marine modelling.

The structure of the Science Programme area has accordingly been shaped so as to more clearly show the leading strategic objectives and explicitly reflect the relevance to the IOC mandate and the needs of its Member States. An adjustment in the

Figure 6.10 Ocean research must be carried on even when the sea is frozen. (a) © Joint Commission for Oceanography and Marine Meteorology (JCOMM); (b) © NOAA.

approach of the IOC to fulfil its obligations and meet the challenges and demands expressed externally and internally is evident, both with respect to coordination and communication.

Technological developments have occurred in parallel with the conceptualization, negotiation and entering into force of overarching international legal instruments. The technological development makes possible ocean observations through satellite remote sensing, moored buoys, drifting floats and other platforms, as well as gathering information from the sub-seabed. All this paired with data communications, high-power computer facilities and dynamic modelling make increasingly useful forecasting possible. The legal instruments both facilitate this and put constraints on the uses of these possibilities. It is only through intergovernmental cooperation and benefit sharing that best use can be made of all the enormous possibilities. The technological development has also made it possible to have citizens directly involved with provision of observations, global networking, outreach and helping in the application of early warning systems. This is evidenced for instance in coral reef research, sea grass bed studies and protected areas of various kinds. This has led to a Citizens Science, in parallel with the Big Science. It provides for studies and capacity development at field level, which is helping to raise public awareness, reaching local authorities as a bottom-up process influencing decision-making. The increasing linking of sciences with the human dimension and the public is seen in this context in the Earth Sciences for Society International Year of Planet Earth, 2007–2009.

The challenges for intergovernmental cooperation include the increasing need for coordination and communication, of and between activities, programmes and organizations, and between and amongst nations. Furthermore there is a strong need for outreach of information and education to communities other than ocean ones to facilitate the application and use of the ocean science results in governance for sustainable development and management of other natural and socio-economic systems such as agriculture, aquaculture, freshwater resources, energy, coastal protection and development, transportation, tourism and recreation. International cooperation in ocean research has been a great success, giving ample returns to Member States and other interests. The many marine issues facing the world today emphasize the need to continue this cooperation, for ocean science to provide reliable and accurate information for the management of marine activities and services and for Member States to ensure the necessary human and other resources are available.

7 The development of ocean climate programmes

ALLYN CLARKE

Allyn Clarke is a Scientist Emeritus at the Bedford Institute of Oceanography, Canada. Since the 1960s, he has participated in and led field programmes focusing on the circulation of the mid and high latitude North Atlantic. He has been a member and leader of many global ocean climate experiments and committees over the past two decades.

Early ocean observations

Before there were oceanographers or intergovernmental commissions, mariners and fishers were observing the changing patterns of winds, weather and currents and passing this information on to family and trusted friends. Popular history credits Prince Henry of Portugal with creating a school of navigation in which the knowledge collected by successive voyages of discovery along the western coasts of Africa was tabulated and used to support ever distant voyages.

While the admirals of the treasure fleets and the pirates and privateers who hunted them knew that the Gulf Stream sped the homeward passage to Spain, this knowledge seemed to be lost to the British postal service to the North American colonies. New England merchants remarked that westbound postal packet boats took weeks longer than eastbound boats. In 1769, Benjamin Franklin, then Deputy Postmaster General, collaborated with Timothy Folger, a Nantucket whaler, to publish a map of the Gulf Stream so that packet boats could avoid its contrary currents on their westward passages. Franklin even made temperature measurements on several transatlantic crossings and identified the temperature front associated with the western edge of the stream.

Much of the early climatology of ocean currents and marine winds was contained in the sailing directions and charts maintained by trading companies (both the

Figure 7.1 Launching an expendable bathythermograph to measure the temperature of the upper layers of the ocean (CSIRO Marine and Atmospheric Research, Australia).

English and the Dutch East India companies, the Hudson Bay Company and the Muscovy Company). This information was jealously guarded in order to preserve the trade monopolies that these companies enjoyed. Many of these documents still exist in archives, museums and rare book collections.

International coordination of ship logs

Modern intergovernmental action with respect to ocean climate began with Lt. Matthew Fontaine Maury, the first superintendent of the US Naval Observatory. Maury had discovered a large archive of ship logs and realized that they formed

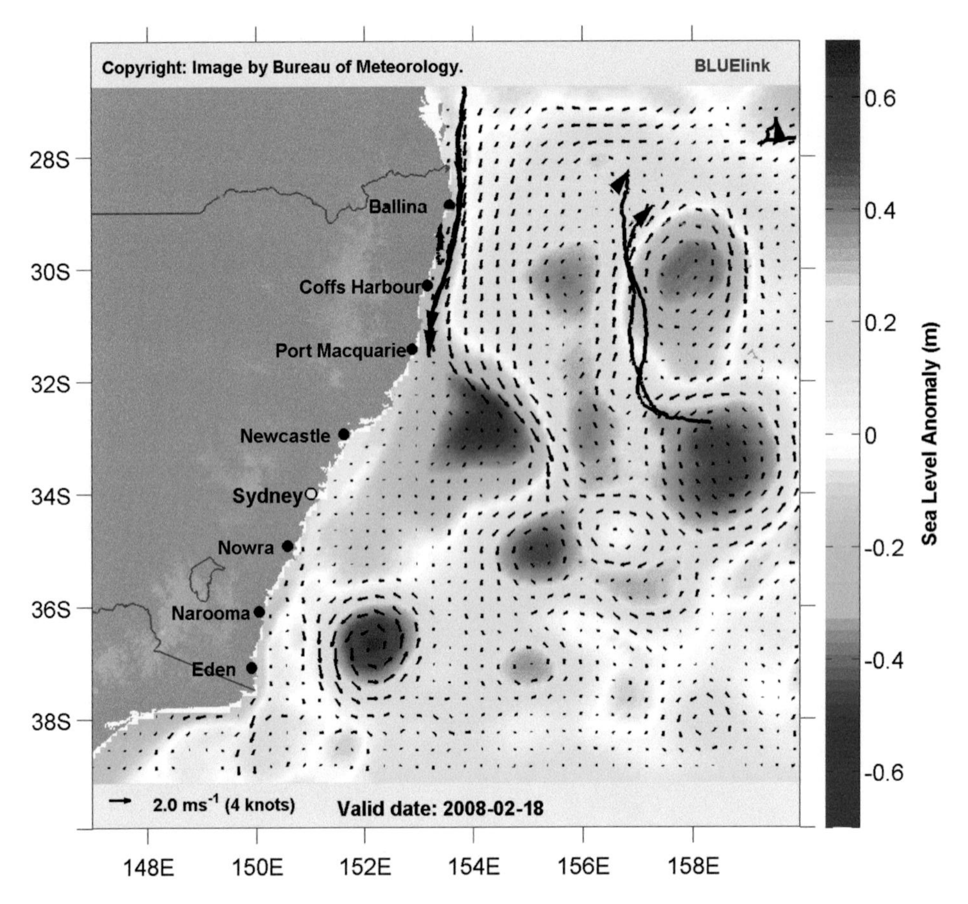

Figure 7.2 Maps of sea level anomalies, generated from satellite altimetry and ocean modelling, show details of the circulation off south-east Australia. © Australian Bureau of Meteorology.

a record of the climatology of marine winds and ocean currents. Using this data, he began creating pilot charts for the world oceans. He soon recognized that this effort would be greatly enhanced with standard methods to observe and record marine weather and surface currents. He established such a system within the US Navy and began to lobby both the US government and his international colleagues to convene an international conference to extend this system to ships of all nations.

While national hydrographic offices supported this idea, rivalry between governments made it difficult to find an appropriate host for the meeting. The logjam was eventually broken in 1853 when King Leopold of Belgium invited nations to send delegates to an International Conference on Oceanography.

Roger Charlier in his 2004 history of this meeting describes its initiation and its accomplishments as:

> An American generated the idea, a Russian buttressed it, an Englishman devised a system, a Belgian convened and chaired it, ten other nations took part in it, and a dozen others chimed in *post facto*.

This conference was the first intergovernmental meeting on ocean climatology. It led to an International Meteorological Congress in Vienna in 1873, and eventually to the founding of the International Meteorological Organization (IMO) in 1905.

The logbook observations were used to create a climatology of marine winds and ocean currents that was delivered to the marine community in the form of monthly pilot charts covering various ocean basins. The collection of data from logbooks has continued to this day and forms the basis of the International Comprehensive Ocean-Atmosphere Data Set (ICOADS). Initially, this international effort collected monthly data from about 12% of the ocean surface in 1855–1857, dropping to about 6% by 1865 before starting a slow rise to 65–70% in the 1980s. The programme has been supplemented in recent years by drifting and moored floats and satellite observations. Intergovernmental direction of this programme continues through the Joint WMO-IOC Technical Commission on Oceanography and Marine Meteorology (JCOMM).

Tides and mean sea level

Mariners and harbour masters were also interested in ocean tides since they governed the times of entry and departure for many ports. The first European tidal observations began in the early seventeenth century, through records of the times and heights of high water. By the second half of the eighteenth century, predictions of future high and low tides were being made and published for all major ports.

Mean sea level was also used as the reference datum for both bathymetric charts and topographic maps. However, sea level was known to change on tidal, meteorological, seasonal and longer time scales. The Permanent Service for Mean Sea Level (PSMSL) was established in Liverpool in 1933 to provide an international repository for sea level data from around the globe. These data have been used to document sea level rise over the past century.

Ocean variability for marine fisheries

Variability of fish stocks in the north-eastern North Atlantic led to the founding of the International Council for the Exploration of the Sea (ICES) in 1902. The Council

Figure 7.3 The Padrao dos Descobrimentos (Monument to the Discoveries) celebrates the Portuguese who took part in the Age of Discovery. It is located in the Belem district of Lisbon, Portugal. The lead figure is Henry the Navigator. © Robert Paul van Beets/Shutterstock.com.

was originally established as a 5-year research effort on the relationships between ocean variability and the availability of fish stocks. A key component of their observational strategy was quarterly hydrographic cruises conducted by specially equipped research vessels from member countries. The term hydrographic related to physical and chemical properties of the ocean which were combined with the biological observations. These observations were coordinated by a central laboratory which established standards for oceanographic observations. The laboratory also created a suite of oceanographic instruments, standard protocols for observations and carried out chemical analysis of sea water samples. A legacy of the central laboratory is

Figure 7.4 This early map of the Gulf Stream location was produced by Benjamin Franklin for the mail service from England, based on information from whaling captain Timothy Folger. This map was rediscovered by P. Richardson (1980), and is remarkably accurate. Image from R. G. Peterson et al. (1996) Early concepts and charts of ocean circulation. *Progress in Oceanography* **37** (1): 1–115.

the Standard Sea Water Service, now overseen by the International Association for the Physical Sciences of the Oceans (IAPSO), which provides the oceanographic community with its standard for salinity. Other regular repeat oceanographic sections were, and continue to be, conducted by various fisheries and marine safety organizations around the globe in addition to ICES.

International ocean data systems

A final component of the modern ocean climate programmes is the global ocean data bases. Up to the 1950s, oceanographic observations were generally preserved in data reports and atlases. Large oceanographic institutions, as well as ICES, had data libraries which consisted mostly of their own data. In the planning for the International Geophysical Year of 1957–58, the International Council of Scientific Unions established a suite of World Data Centres, each focused on a particular geophysical discipline. World Data Centres for Oceanography were established in both the USA and the Soviet Union and these data centres continue to this day.

The IOC recognized that data archiving and exchange were important components of all ocean programmes and established a Working Group on Oceanographic Data Exchange at its first assembly in 1961. This Working Group has evolved into the

International Oceanographic Data and Information Exchange programme of the IOC. The task of this Working Committee is to improve the quality and quantity of oceanographic data exchange and archiving.

Global Atmospheric Research Programme (GARP)

Up to the 1970s, oceanographers and the IOC were interested in ocean climatology: the description of the ocean's mean state and an understanding of its modes of variability. With the arrival of GARP, the oceanographic community was challenged to see the ocean as a component in a global climate system. GARP wasn't designed as a climate programme; it was a programme that sought to extend the capability of weather forecast models to the global domain and to periods of 7–60 days. The extension of numerical models to the global domain required atmospheric observations over the whole ocean, including the broad expanse of mostly southern oceans in which few ships sailed. The extension of the forecast times from 1–4 days to weeks required better knowledge of the heating and cooling of the ocean surface layer and its interaction with the atmosphere.

Atmospheric scientists had recognized the importance of the tropical ocean to global and regional atmospheric dynamics. Among the first GARP experiments was the GARP Atlantic Tropical Experiment (GATE) in 1974. This experiment was designed to investigate atmospheric deep convection processes over the tropical ocean. The experiment was originally planned for the western Pacific where the many islands and atolls would provide sites for a meteorological observation network.

Difficulties in gaining access to key sites led to moving the experiment to the Atlantic, where the lack of island sites required the establishment of a meteorological observational network based on ships, many of which were to be research vessels. Faced with the reality that a large number of the world's deep sea research vessels were committed to 3 months in the tropical Atlantic, SCOR established Working Group 43 to coordinate an oceanographic research programme that could complement the atmospheric programme. This working group was chaired by Gerold Siedler of the University of Kiel. Several of the members, for example John Woods, George Phillander and George Needler, went on to play leading roles in the climate research programmes of the 1990s. Much that was already known about the ocean dynamics of the equatorial Atlantic had been obtained during the EQUALANTS surveys of the early 1960s. These surveys covering the entire tropical Atlantic were among the first international field programmes planned and coordinated through the IOC. The information and collaborative research from these earlier programmes allowed

Figure 7.5 The development of the Argo float revolutionized ocean observations. © JCOMM.

the oceanographic community to quickly plan an effective scientific plan for GATE.

The experiment was conducted during the summer of 1974. Research vessels stationed across the tropical Atlantic carried out a coordinated programme of meteorological and oceanographic observations. A key contribution of GATE was the creation of a general self-describing data exchange and archiving format for the full suite of oceanographic observations. The IOC supported GATE through its subsidiary bodies on data exchange. During the analysis phase of GATE, the IOC took responsibility for the production and publication of a three-volume Atlas based on the ocean observations in GATE.

GATE was just one component of GARP and as it further developed its programmes, it sought further collaboration with the oceanographic community. Robert (Bob) Stewart, a Canadian oceanographer interested in air–sea exchanges as well as general problems of geophysical fluid dynamics, was a member of the Joint Organizing Committee for GARP and argued forcibly for the inclusion of ocean research in its programme. Both the IOC and SCOR responded to the challenge. In 1978, the IOC formed an *ad hoc* group of experts led by Hans Roll to investigate potential oceanographic programmes under GARP. This working group focused on the intergovernmental needs of such programmes in areas of data exchange, standards and training, while a SCOR Working Group chaired by Henry Stommel from the USA focused on the scientific design.

Both working groups concentrated on the First GARP Global Experiment (FGGE). Like GATE, FGGE was primarily an atmospheric programme. It sought to integrate new satellite observations, standard and novel meteorological observations and numerical models in order to describe and forecast global weather. The programme began in 1978, during which new instruments and observing systems were installed and tested. 1979 was the operational year with intensive observations during January–February and May–June.

Oceanographers were particularly interested in two aspects of the FGGE field programme. FGGE plans called for 50 vessels to be stationed in the tropics to measure the wind fields and complementary oceanographic programmes were designed by SCOR for all three tropical oceans.

The second aspect was the planned deployment of a large number of surface drifters in the Southern Ocean. These drifters were tracked by satellite using the new Service Argos system. Their primary task was to report air pressure and sea surface temperature over ocean areas that had little commercial shipping. The floats were equipped with drogues to provide surface current information to the oceanographic community.

The IOC's principal contribution to FGGE was through its support of the Integrated Global Ocean Station System (IGOSS), jointly supported by the IOC and WMO since 1975. IGOSS is an operational system to collect, transmit and distribute oceanographic data in real and near real-time and produce and distribute ocean information products based on these data. Data transfer and management under IODE and IGOSS were important IOC contributions to FGGE. The data centre for drifting buoy data, established in Canada on behalf of the IOC, continues to this day as a Responsible National Oceanographic Data Centre.

One of the core activities of the IOC involves its regional bodies. These bodies have provided forums for local scientists to plan and coordinate programmes focusing on regional issues. Before El Niño was recognized as a global climate signal, it was known as a significant ocean variability phenomenon in the eastern tropical Pacific. The Permanent Commission for the south-east Pacific (CPPS) joined with the WMO and the IOC to help coordinate FGGE activities in this region.

Pilot Ocean Monitoring Programme

During the 1970s, the oceanographic community was working to develop realistic ways to monitor ocean variability. This long-standing objective had been the primary oceanographic goal which led to the founding of the ICES in 1902. Experiments like MODE and POLYMODE had begun to reveal the strength of mesoscale variability of the deep ocean. Sea surface temperature maps from space exposed the complexity of the ocean surface structure and dynamics. GARP with its focus on extended weather forecasting was primarily interested in the upper mixed layer of the ocean. The oceanographic community realized that the ocean's role in climate was more than simply a surface boundary condition for the atmosphere. And again SCOR responded with a Working Group to report on the 'Influence of the Ocean on Climate'.

Figure 7.6 SEASAT, the short-lived but successful satellite demonstration of sea level measurement from space. © NASA.

In November 1978, the Joint Organizing Committee for GARP held a study meeting with SCOR in Kiel on 'The role of the Ocean in the Global Heat Budget'. This led to a meeting, in Miami in October 1979, to plan for a Pilot Ocean Monitoring Study (POMS) that was again jointly organized by SCOR and the JOC and co-sponsored by the IOC. This meeting recommended the development of what eventually became the oceanographic component of the World Climate Research Programme (WCRP).

The First World Climate Conference and the establishment of the WCRP and CCCO

Much of the information concerning the early development of ocean climate programmes is found in reports of the various sponsoring organizations and steering committees. Most of these reports are out of print and are only found in leading oceanographic libraries. The First World Climate Conference was hosted by the WMO in Geneva in February 1978. This meeting was primarily a scientific meeting, but also demonstrated support for the establishment of a World Climate Programme. Most of the scientific papers dealt with the occurrence and impact of droughts, floods and other climate variations. The rising level of carbon dioxide was a developing concern; however, the prevailing view was that the associated global warming would take place slowly. A higher priority was that rising population levels had increased the vulnerability of human populations to the impacts of natural climate variability. The proposed World Climate Programme was intended to develop the knowledge, numerical models and supporting observations to support timely and useful forecasts of climate variations in order that appropriate mitigation measures could be put in place.

In 1980 the World Climate Programme (WCP) was established under the co-sponsorship of the WMO and ICSU and consisted of four programmes covering data, applications, impact studies and research respectively. The first two were managed by the WMO, the impact studies by the UNEP and the research jointly by the WMO and ICSU. SCOR played an important role in providing the oceanic expertise that ICSU needed to exercise its sponsorship of the WCRP.

SCOR established the related Committee on Climate Change and the Ocean (CCCO) in November 1978 and IOC accepted the invitation to co-sponsor the committee and subsequently hosted the Secretariat of the CCCO in its office in Paris. The committee had a mandate to assess the role of the ocean in climate change and conversely the effect of climatic changes within the ocean and to identify research problems, their facilitation and solution. In addition the Committee was charged with keeping ocean science efforts under review, to advise on capacity building and assistance requirements and to foster collaboration and cooperation in ocean climate research.

The first meeting was held in October 1979. The membership of this committee was well chosen to guide its broad mandate. The chairman, Roger Revelle, had a distinguished career as an oceanographer and a builder of oceanographic institutions both in the USA and internationally. Through the 1970s, his scientific publications built the case that rising carbon dioxide levels in the atmosphere was an important climate issue. Other members of the committee were also scientific leaders in their fields; many occupied management positions within their national oceanographic institutions and were well placed to promote climate research programmes in their own countries. Tomio Asai worked on observing and modelling exchanges between sea and atmosphere and their impact on the atmosphere, Leoid M. Brekhovskikh was an expert in ocean acoustics, Kirk Bryan was a pioneer in ocean modelling, John Woods was a scientific leader in mixed layer processes, Adrian Gill was a leading theoretician with special interest in equatorial dynamics, Jörn Thiede was a marine geophysicist studying paleoclimate from marine sediments, Alan Longhurst was a marine ecologist studying the global distributions of marine ecosystems and Robert W. Stewart was a geophysical fluid dynamicist with special expertise in air–sea exchange processes.

At its first meeting, the CCCO established four panels covering areas of special research interest. These areas were respectively: the theory and modelling of ocean dynamics and climate, marine ecological climate studies and the evaluation of long time series of oceanography, marine biology and fisheries observations; paleoclimatology over the past few centuries to the past few millennia; and sea ice variability, including linkages between sea ice processes and the formation of deep waters in high latitude regions.

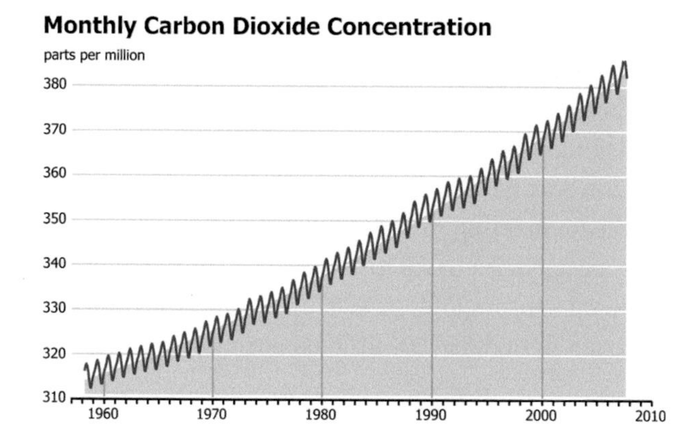

Figure 7.7 The 'Keeling curve' of carbon dioxide measurements taken at mount Mauna Loa, Hawaii over nearly half a century, showing the long-term rise and annual fluctuations. © IOC/UNESCO.

The meeting also reviewed the report of the POMS meeting in Miami and recommended a careful examination of the results of the existing time series and other observations before undertaking significant expansion of these systems. The members cautioned that existing ocean time series were few and should not be terminated or altered until confident that the altered observations will provide a seamless continuation of the original.

Development of the climate research programmes of 1985–2010

The POMS meeting in Miami also identified three potential foci for ocean climate research programmes.

The first was to observe the north–south heat transport carried by the combined ocean and atmosphere over an entire ocean basin in order to describe the processes involved. The CCCO and JSC jointly commissioned a study under the leadership of Fred W. Dobson to investigate the feasibility of such a proposal. The North Atlantic basin was chosen since it was well observed by both research vessels and commercial shipping; it was the narrowest of all the ocean basins and expected to produce highly visible observations arising from its vigorous overturning circulation. A study team of oceanographers and meteorologists used existing data to estimate all of the terms in the atmospheric and oceanic heat transport and came to the conclusion that the uncertainties in the various terms were too large to achieve the desired results over the basin and it was decided not to pursue the project.

The second potential project was to design a programme to determine global ocean circulation using a mix of hydrographic stations and satellite altimetry. This idea was first proposed by Carl Wunsch and would require the involvement of research vessels and a dedicated satellite commitment. The SEASAT satellite had succumbed to a massive power failure 3 months into its mission in 1978; however, its altimeter had demonstrated variations in sea surface height could be observed with sufficient precision to observe changes in ocean currents. An extended satellite altimeter mission was needed for this programme to go forward. The JSC and CCCO called a meeting in January 1981 with representatives of the various space agencies to discuss the scientific needs of the climate programmes for space based sensors and, in 1981, the Centre National d'Études Spatiales (CNES) approved flying the POSEIDON altimeter on their SPOT satellite.

Concurrently, the CCCO asked Francis Bretherton to lead a study group on the World Ocean Circulation Experiment (WOCE). This study group developed a strategic plan for WOCE and presented it to the Tokyo Conference on Large-Scale Oceanographic Experiments in the WCRP in May 1982. The conference endorsed the concept and, in 1983, the JSC/CCCO Scientific Steering Group for WOCE was established under Carl Wunsch. The plans for WOCE were further developed under an International WOCE Planning Office led by George Needler, greatly aided by the US WOCE Office led by Worth Nowlin. A WOCE Scientific plan and an Implementation Plan were developed and presented to the WOCE Scientific Conference hosted by the IOC in December 1988 and the programme was approved to begin its 7-year field phase at the beginning of 1990.

The third initial focus was the need for long-term monitoring of ocean variability. New instruments for both *in situ* and spaced base observations of the ocean offered some hope for new efficiencies. However, there was also a concern that new programmes would replace the older programmes resulting in abrupt discontinuities in the climatic records. A small study group on the Ocean Observing System Development Programme (OOSDP) was established under Klaus Voigt. This group served as an interface between the scientific needs defined by the new ocean climate programmes and the operational services within the IOC and WMO that provided data and programme management.

At their initial meeting, the CCCO did not discuss any programmes focused on tropical processes. However, a community of tropical meteorologists and oceanographers were carrying out observational studies in both the Atlantic and Pacific. Their publications were documenting links between anomalies in tropical SST and sea level with anomalies in tropical wind patterns and climate variability at higher latitudes. The CCCO and JSC jointly established a study group in 1982 on these issues, with Adrian Gill as chair. By November 1982, the strongest El Niño ever observed was underway. The study group quickly evolved into the Scientific Steering Group for the Tropical Ocean Global Atmosphere (TOGA) programme. The TOGA Scientific plan was presented to an International TOGA

Figure 7.8 Maintaining the buoys providing permanent monitoring of the El Niño events is an expensive but extremely important task. © NOAA.

Scientific Conference hosted by the IOC in Paris in September 1984. This plan was accepted by the conference and TOGA was launched in January 1985. One major TOGA accomplishment was the establishment of a basin wide mooring array for the monitoring of El Niño in the tropical Pacific under the leadership of Stan Hayes. This array has been expanded into the Indian Ocean and to higher latitude and continues to the present as the TOA/TRITON array.

IOC data management services contributed greatly to both TOGA and WOCE; however, it took a great deal of effort to improve their timeliness and quality. Initial investigations of the IGOSS system revealed that many oceanographic messages were being lost by the Global Telecommunication Network (GTS). WOCE also demonstrated that high-quality hydrographic and tracer data could be exchanged more rapidly than had been the norm to date. Jim Crease played an important role in this integration between the IOC's data services and the ocean climate research programmes and was an active participant in both the scientific planning for WOCE and the intergovernmental management of IODE and IGOOS.

The oceanographic community had a long interest in the carbon cycle in the oceans. The global distribution of alkalinity and total carbonate had been observed during the GEOSECS programme of 1972 to 1978. These observations had been repeated in many ocean basins in the 1980s by various transient tracer expeditions. It was recognized that the oceans constituted the largest reservoir of carbon dioxide on the planet other than the deposits of sedimentary rocks. Hence the air–sea exchange of carbon dioxide and the subsequent exchanges between the upper ocean and the deep waters and marine sediments are important climate processes.

Since WOCE was already planning to collect water samples for transient tracer investigations of deep water circulation patterns, it agreed to support an ocean carbon programme by providing water samples and scientific berths on all WOCE sections. Scientists in a number of nations, led by marine biologists, were interested in investigating the processes that remove carbon from the ocean surface layer. They were particularly interested in the role that the biosphere played in these transfers. SCOR and ICSU co-sponsored a meeting of experts in Paris in February 1987. This meeting established a scientific strategy and organizational structure for the Joint Global Ocean Fluxes Study (JGOFS). SCOR formed an International Planning Committee for JGOFS which began meeting in 1988. JGOFS became one of the initial core projects of the International Geosphere Biosphere Programme (IGBP) on Global Change, established by ICSU in 1987.

In 1991, the CCCO was disbanded in favour of a closer IOC relationship with the WCRP. Three oceanographers were added to the JSC to provide the appropriate oceanographic expertise and in 1993 the IOC became a co-sponsor of the WCRP.

As TOGA(1995), WOCE (2002) and JGOFS approached their termination dates, the scientific community began planning to transfer some programmes as operational programmes under the Global Ocean Observing System (GOOS) and the Global Climate Observing System (GCOS) and others to the succeeding research programmes such as the Climate Variability and Predictability Study (CLIVAR) and Surface Ocean – Lower Atmosphere Study (SOLAS).

The future – operational oceanography

The CCCO and JSC established the Ocean Observing System Development Panel (OOSDP) in 1990 under the leadership of Worth Nowlin. The Second World Climate Conference, held in Geneva in November 1990, also called for the establishment of an effective Global Climate Observing System (GCOS) that would provide the observational basis for predicting climate variability and detecting climate change. In addition, during this same period, the IOC with the support of the WMO and ICSU were preparing a planning structure for a Global Ocean Observing System (GOOS). These efforts were complementary and all bodies decided that the scientific and technical plan being developed by the OOSDP would be the ocean component of GCOS and the climate component of GOOS.

GOOS is co-sponsored by the IOC, UNEP, WMO and ICSU and managed by the IOC. The resources for GOOS are supplied by nations; nations have been reluctant to allocate new operational funds to continue programmes that were initiated with research funds. However, an operational El Niño observing system in the tropical Pacific and eastern Indian Ocean does exist. As a consequence, it is hoped that future ocean scientists interested in doing climate research will be able to spend less of their time designing and organizing observational programmes and more in the analysis of data provided by the IOC through GOOS.

8

The IOC's International Bathymetric Chart Series: a programme facing extermination?

RON MACNAB AND DMITRI TRAVIN

Ron Macnab is a retired marine geophysicist whose research career included participation in several major data compilation projects. He is chairman of the IASC/IOC/IHO Editorial Board for the International Bathymetric Chart of the Arctic Ocean (IBCAO).

Dmitri Travin is a retired Captain of the Russian Navy whose professional career included participation in numerous hydrographic surveys around the world. At present, he is the Professional Assistant for the Ocean Mapping Programme of the Intergovernmental Oceanographic Commission of UNESCO.

Accurate and detailed descriptions of depth and morphology are essential to the advancement of marine science near the world's coastlines, over adjacent continental shelves and in the deep ocean. In recognition of the broad benefits to be gained from reliable portrayals of the global seabed, a chart production programme was proposed over 30 years ago as a multinational undertaking under the auspices of the Intergovernmental Oceanographic Commission.

Launched in 1974, the IOC's International Bathymetric Charts programme provided a framework for compiling existing soundings to produce detailed maps of sea floor relief in selected regions around the world. In due course, an Ocean Mapping

Programme was established by the IOC with a view to supporting and coordinating these compilation activities.

Conceived in an era when marine cartography was still largely a manual and labour-intensive process, the IBCs have evolved to reflect advances in the technology for visualizing and manipulating bathymetric information in digital form. In spite of its historic accomplishments and its recent developments, the IBC operation now faces an uncertain future in light of reduced financial support from the IOC and the reported decision to downgrade the Ocean Mapping Programme.

Why accurate maps of the seabed are important, and why they aren't being produced

Key among the research concerns of the 1970s was the influence of bottom topography on the movement of oceanic water masses: scientists were beginning to model this movement with tools that incorporated boundary conditions in the form of digital bathymetric grids which described seabed relief in a highly generalized fashion. This activity continues in the present day, but the modelling tools at the disposal of the modern researcher are much more sophisticated, and the requirements for detailed descriptions of seabed relief in both shallow and deep areas are correspondingly more stringent.

To illustrate just one modelling application, the Boxing Day Tsunami of 2004 emphasized the need to predict the propagation of tsunami to near and distant shores, and to forecast the likelihood of catastrophic flooding along vulnerable coastlines. The reliability of these forecasts is severely restricted where nearshore, coastal and deep water bathymetry remain poorly known.

In coastal regions, the IOC has mounted a partial response to this knowledge gap by initiating programmes for building the capacity of coastal states to perform nearshore mapping in areas that are prone to tsunami damage. However there remains the significant problem of mapping the topography of the deep seabed in order to predict accurately how tsunami will propagate. Deep ocean ridges, for example, have been found to act as wave guides that focus tsunami energy – therefore better descriptions of these features should yield more reliable predictions concerning the intensity and direction of tsunami propagation.

Many if not most deep-sea regions have never been systematically mapped, and it will likely be years before they are properly surveyed using the required acoustic technology. Cost is a significant impediment to this type of mapping, because for the most part it involves the deployment of high-priced technology on board vessels that are slow and expensive to operate. This truth needs to be placed in perspective: it has been estimated that the cost of a global multiship ocean mapping programme would be about equal to the cost of mobilizing a single exploratory mission to the outer reaches of our solar system.

While observations of satellite altimetry can be used to derive approximate descriptions of the seabed, the values so produced lack the accuracy and the resolution that are necessary for detailed investigations

An additional and considerable disincentive to launching any acoustic mapping programme in deep water is the fact that most of the world's deep-sea areas lie beyond the jurisdiction of any coastal state: mapping such areas is often difficult for government or commercial agencies to justify because it's not clear who will benefit the most from the information. In this vein, ocean mapping has tended to suffer because it's all too often perceived as the gratification of scientific curiosity, when in fact the observations are often applicable to myriad practical uses that transcend narrow national interests. Tsunami prediction is one such application.

Other applications among many that would benefit from improved maps of the seabed include: resolving the influence of seabed morphology and composition on the development of marine habitats; studying hydrothermal vents with their unique benthic communities and mineralizations; and investigating topographically induced upwelling. A better comprehension of these three phenomena would significantly advance efforts to understand and preserve biodiversity in the world ocean.

At one time, it was anticipated that the global reach of Article 76 of the United Nations Convention on the Law of the Sea (UNCLOS) would generate significant contributions of new data acquired during surveys undertaken by coastal states for determining the outer limits of their extended continental shelves. While Article 76 has motivated deep-water mapping missions in many parts of the world, the expected benefits may not be as substantial as initially envisaged: in keeping with their specific objectives, many of these specialized survey missions restrict their focus to locating and describing just two features on the seabed, i.e. the foot of the slope and the 2500 m isobath. UNCLOS cruises may therefore miss the opportunity of developing more complete views of the seabed through broader survey programmes that encompass adjacent areas. An added uncertainty with UNCLOS data sets could be the prospective tendency of some states to withhold their new observations, rather than to place them in the public domain where they could be put to good use for map compilation and scientific research.

Bathymetric compilations as preliminaries to ocean surveys: a role for the IOC

Pending the implementation of a global ocean mapping programme, agencies and institutions could respond constructively to the situation by promoting and supporting compilation projects that assembled and rationalized existing deep-sea acoustic data in order to produce the best maps possible. By developing improved portrayals of the seabed, these initiatives would not only improve our current view of the seafloor but they would also provide extremely useful tools for planning future shipboard operations.

In this vein, the IOC could play a major role by underlining to its member states, and to the oceanographic community at large, the benefits of bathymetric compilations as essential precursors to systematic ocean surveys. At the same time, it could bring its influence to bear on member states by encouraging them to enable their respective agencies and institutions to engage in such activities. In pursuing this goal, the IOC could also recruit a sister organization that concerns itself with bathymetric mapping: the International Hydrographic Organization (IHO), with a membership that comprises many if not most of the world's government-sponsored hydrographic agencies.

As a subsidiary body of the IOC and the IHO, the organization known as the General Bathymetric Chart of the Oceans (GEBCO) could provide a means for organizing and coordinating an effective global compilation project, advocating all the while for an international programme of coherent and systematic seabed surveys on a worldwide basis. GEBCO is unique in that it occupies an intermediate position between the ocean mapping community and the member states of its parent bodies IOC and IHO. Hence it seems ideally suited to provide a communications link between ocean mappers and governments, and to broker a broadly supported arrangement that would enable a successful approach to bathymetric compilations and surveys. Regrettably, GEBCO has been less than effective in rising to this challenge: as an organization that is essentially unfunded, it appears to be engaged in a perpetual struggle to focus the interests, the energies and the enthusiasms of volunteer members who possess diverging agendas.

In promoting bathymetric compilations and ocean mapping initiatives to its member states, the IOC would be operating in a fashion consistent with its current roles in advocating and supporting global operations that require the willing participation of many international partners. Examples of such activities that come readily to mind are: Capacity Development; the Global Ocean Observing System (GOOS); and the International Oceanographic Data and Information Exchange (IODE). These and other operations require the IOC to provide technical and administrative leadership, and to assume some responsibility for ensuring the availability of adequate funding.

The early (analogue) history of the IBCs

The first IBC series consisted of the 10 1:1 000 000 sheets of the International Bathymetric Chart of the Mediterranean (IBCM) and its Geological–Geophysical series. Printed in 1981 by the Head Department of Navigation and Oceanography of the Soviet Union, these full-colour charts featured hand-drawn depth contours. Geological and geophysical overlays followed over the next two decades.

As the first in the series, the IBCM provided a platform for the development of techniques and specifications that would be employed in subsequent productions. Reflecting the state of cartographic technology at the time, heavy reliance was placed on manual compilation and construction methods, with projects divided into manageable portions that were assigned to separate production teams in order to spread the workload. The process began with plotting sheets that covered the Mediterranean and Black Seas at a scale of 1:250 000, and which provided a base for compiling point soundings. These were subsequently contoured and consolidated on 1:1 000 000 sheets for publication.

Following the IBCM model, the IOC initiated IBC projects for the production of bathymetric contour maps in five different regions: the Caribbean Sea and the Gulf of Mexico (IBCCA); the Central Eastern Atlantic Ocean (IBCEA); the Western Indian Ocean (IBCWIO); the Western Pacific Ocean (IBCWP); and the South-east Pacific Ocean (IBCSEP). The responsibility for developing each IBC was assigned to an Editorial Board, essentially a team of technical specialists who had frequent exchanges to discuss problems and to coordinate activities.

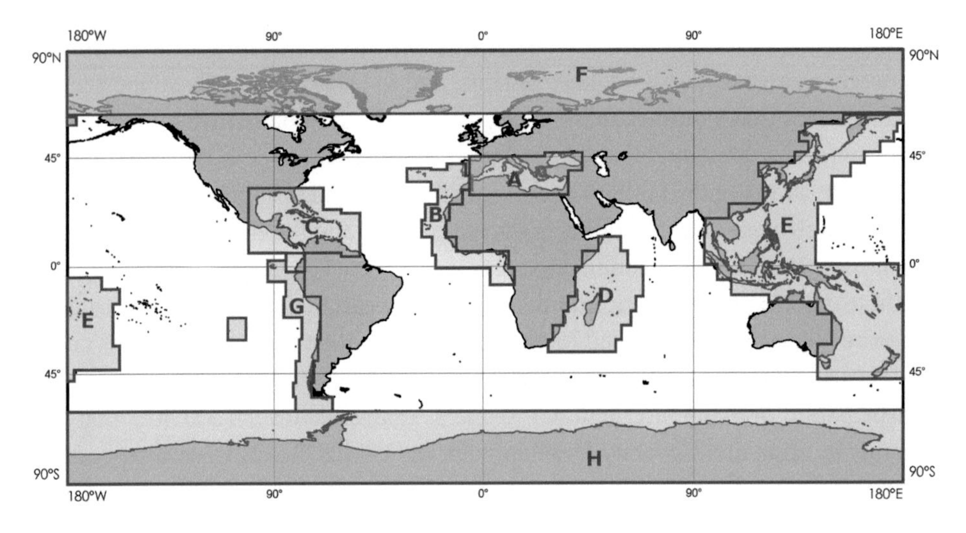

Figure 8.1 Locations and extent of current International Bathymetric Charts (IBCs). A: IBCM (Mediterranean and Black Seas); B: IBCEA (East Central Atlantic Ocean); C: IBCCA (Caribbean Sea and Gulf of Mexico); D: IBCWIO (Western Indian Ocean); E: IBCWP (Western Pacific Ocean); F: IBCAO (Arctic Ocean); G: IBCSEP (South-East Pacific Ocean); H: IBCSO (Southern Ocean). Not shown is the outline for IBCNA (North Atlantic Ocean), which extends from the Equator to the Arctic Circle. IBCAO, IBCSO and IBCNA are all-digital products, with depth grids as their primary output. For the most part, the remaining IBCs portray ocean depths in the traditional form of printed contour maps, each IBC comprising a series of 1:1 000 000 edge-matched sheets. Figure by Ron McNab.

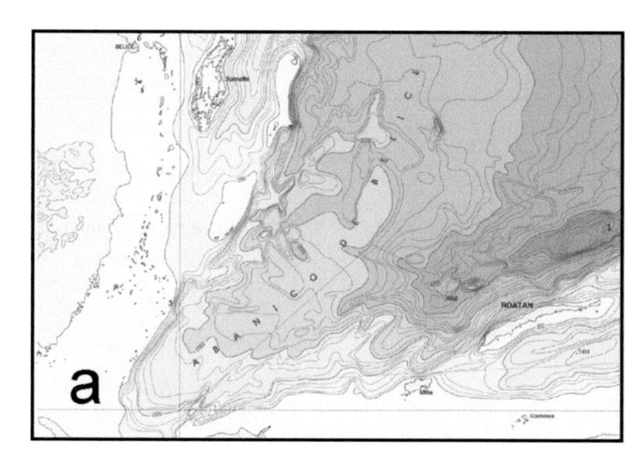

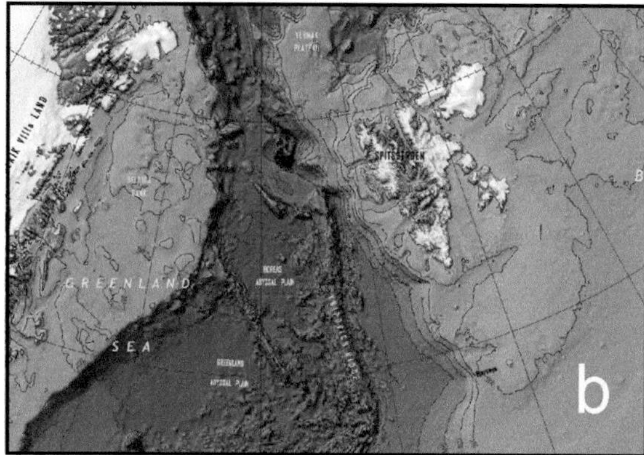

Figure 8.2 (a) Extract from the 'original style' 1:1 000 000 Sheet 1.06 of the International Bathymetric Chart of the Caribbean and Gulf of Mexico (IBCCA), showing the portrayal of ocean depths in the Gulf of Honduras by means of isobaths. (Image courtesy of Jose Frias Salazar, Instituto Nacional de Estadística Geografía e Informática, Mexico). (b) Extract from a 1:6 000 000 version of the 'new style' International Bathymetric Chart of the Arctic Ocean (IBCAO), showing the method of portraying ocean depths in Fram Strait through a shaded relief representation, with an overlay of isobaths. © US National Geophysical Data Center.

Figure 8.1 portrays the locations and extents of these IBCs. Figure 8.2a is an extract from IBCCA Sheet 1.06 in the inner Gulf of Honduras, showing how depths in the early IBCs were presented in the form of 'traditional' isobaths. While the final IBCCA retained the conventional presentation format that displayed isobaths at standard depth intervals, its construction was based very much on the application of digital methodologies, with little reliance on the manual techniques that characterized earlier IBCs. In that sense, the IBCCA represented a transitional phase in the IBC programme, moving away from reliance on traditional cartography and towards the adoption of modern GIS tools.

Table 8.1 lists the URLs of websites where detailed technical and administrative information (including participating agencies, project teams and ordering information) may be obtained for each IBC. All but three charts comprise an agglomeration of 1:1 000 000 edge-matched sheets. It will be noted that not all sheets in all series have been completed: as can be seen in Table 8.1, a number of IBCs remain as works in progress pending the resolution of technical and administrative issues, notably in the West Indian and West Pacific Oceans.

Challenges in project management

Some of the early IBC series demonstrated the feasibility of the IBCM model, whereas others posed significant project management challenges: with an overall scheme consisting of six separate project areas that encompassed about 175 separate 1:1 000 000

Table 8.1

Series	Location	Year begun	Constituent Sheets			Project Website
			Total	In prep	Done	
IBCM	Mediterranean and Black Seas	1974	10	–	10	http://www.ngdc.noaa.gov/mgg/ibcm/ibcm.html
IBCCEA	Central Eastern Atlantic Ocean	1988	12	4	8	http://www.ngdc.noaa.gov/mgg/ibcea/start_e.htm
IBCCA	Caribbean Sea and Gulf of Mexico	1983	17	–	17	http://www.ngdc.noaa.gov/mgg/ibcca/ibcca.html
IBCWIO	Western Indian Ocean	1982	21	2	5	http://www.ngdc.noaa.gov/mgg/ibcwio/ibcwio.html
IBCWP	Western Pacific Ocean	1993	102	25	–	
IBCAO	Arctic Ocean	1997	1	–	1	http://www.ngdc.noaa.gov/mgg/bathymetry/arctic/arctic.html
IBCSEP	South-east Pacific Ocean	2001	12	4	2	http://www.inocar.mil.ec/IBCSEP/english/index.html
IBCSO	Southern Ocean	2006	1	1	–	http://www.ibcso.org
IBCNA	North Atlantic Ocean	2006	1	1	–	http://www.geo.su.se/geology/ibcna

This table itemizes the salient facts for each of the current International Bathymetric Chart (IBC) series: the oceanic region that is encompassed by the IBC; the year in which it was launched; the number of its constituent sheets and their status (in preparation or complete); and a project website where detailed technical and administrative information may be obtained, such as data sources, construction procedures, names of participating agencies, members of project teams, ordering information, etc.

sheets, project managers and participants were faced with the necessity of assembling, absorbing and manipulating information pertaining to many sheet areas, and of managing the overall process so that seamless outputs were generated on time and according to specifications. This approach had several disadvantages:

- It was often difficult to monitor progress over so many fronts in order to identify problems and to resolve them in a timely fashion.
- Project areas were divided arbitrarily, promoting the fragmentation of data sets that should have otherwise remained intact, and abetting a form of 'tunnel vision' within some project subgroups.

Figure 8.3 Chilean chart checking. Producing and maintaining accurate information on ocean and coastal bathymetry is a specialized and skilful operation. © IHB/IHO.

- There was a significant cost in production and communication overhead, given the necessity of matching the contents of adjoining sheets, and of sharing information among numerous project teams.
- There was a strong potential for duplication of effort when work was pursued independently in overlapping areas.
- There were prospects of incompatible products arising from the use of different data sets.

The transition to digital methods and a proposed restructuring of the IBC scheme

In 1997, a new approach to constructing IBCs was adopted with the launch of the International Bathymetric Chart of the Arctic Ocean (IBCAO), which set out to produce a seamless description of seabed relief north of 64°N. Unlike previous IBCs, IBCAO was developed in a totally digital fashion, without any of the cartographic constraints imposed by the scale and fragmentation of intermediate plotting sheets, or

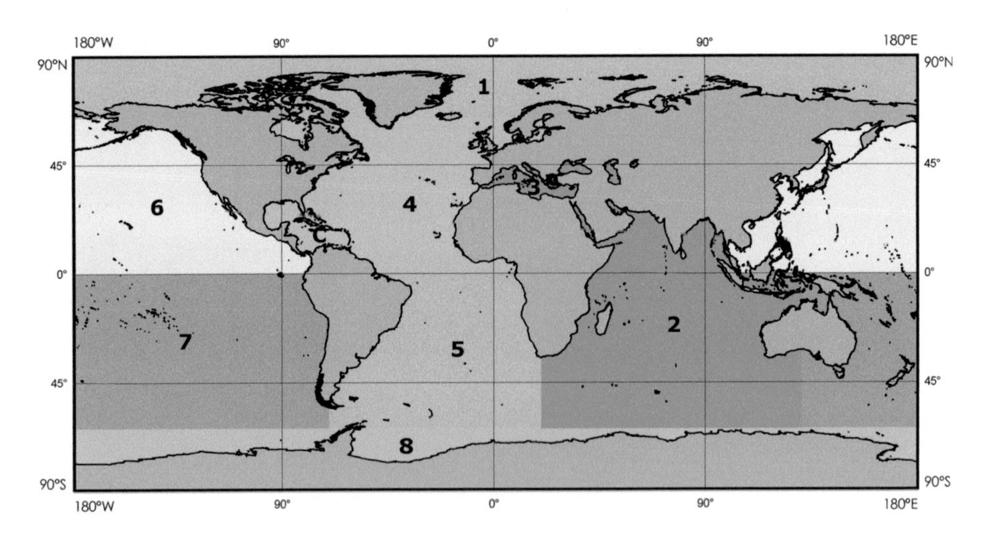

Figure 8.4 Suggested layout for proposed all-digital re-structuring of International Bathymetric Charts (IBCs).1: IBCAO (Arctic Ocean); 2: IBCIO (Indian Ocean); 3: IBCM (Mediterranean Sea); 4: IBCNA (North Atlantic Ocean); 5: IBCSA (South Atlantic Ocean); 6: IBCNP (North Pacific Ocean); 7: IBCSP (South Pacific Ocean); 8: IBCSO (Southern Ocean). Each IBC will provide a framework for compiling soundings in digital form, which will then be combined into a grid of depth values. One of these IBCs (IBCAO) already exists and matches this proposed configuration. Two others (IBCSO and IBCNA) are under construction in conformity with the layout shown here. A fourth (IBCIO) could be readily developed from existing data. A fifth (IBCSA) has been promoted, but has not yet been acted upon. Figure by Ron McNab.

by the projection of the final published product. IBCAO also set out deliberately to encompass the bathymetry of an entire oceanic region, extending from the nearshore to abyssal depths.

Further differentiating it from previous IBCs, the primary output of IBCAO was not a contour map, but a digital grid of depth values that could be manipulated effectively in a variety of visualization and computational processes. A 1:6 000 000 map produced from this grid portrayed the seafloor in shaded relief form (Figure 8.2b), as opposed to the linear isobaths of earlier maps; this representation was quickly accepted by the polar research community as a standard representation of the Arctic seabed. This approach had several advantages:

- A single project area resulted in a simpler, leaner management structure, with less production and communication overhead, permitting the project team to maintain a view of the 'big picture'.
- The project area was naturally and geographically integrated, so major features could be defined coherently without concern for intervening sheet limits or fragmented data sets.

- There was no scope for duplication of effort between overlapping project areas.
- A single database facilitated the rationalization and adjustment of observations, along with the application of uniform specifications throughout the project area.

Following the success of IBCAO, the International Bathymetric Chart of the Southern Ocean (IBCSO) was initiated along similar lines in 2006 with a view to portraying ocean depths south of 60°S. In the same year, the International Bathymetric Chart of the North Atlantic (IBCNA) was launched to produce a description of ocean depths from the Equator north to the Arctic Circle; IBCNA still awaits formal IOC approval, but the assembly and rationalization of soundings from the map area have begun in earnest.

There has also been a suggestion for the development of the International Bathymetric Chart of the South Atlantic (IBCSA) for describing bathymetry between the Equator and the Antarctic Circle, but that proposition has yet to be acted upon. At the same time, there has been some discussion concerning the feasibility of constructing the International Bathymetric Chart of the Indian Ocean (IBCIO), derived from a significant body of international data sets that have been accumulated by investigators at the Scripps Institution of Oceanography.

The practicality of IBCAO, IBCSO and IBCNA has spawned proposals for a reorganization and simplification of the IBC project structure so that it encompasses the entire world ocean in just eight project areas, each corresponding to a region defined by a major ocean and its marginal seas. The proposed project layout is shown in Figure 8.4. If this proposal were accepted, the existing IBC series could be easily integrated into the new, all-digital project structure.

Future prospects of the IBC Programme

The IBC Programme has achieved success in some quarters, but clearly it has fared poorly in others, due largely to a lack of financial and institutional support that has led to a withering of essential monetary, technical and human resources. For whatever reason, national agencies which supported regional IBC projects at one time no longer appear to have the will or the resources to continue sponsoring these activities. By their very nature, national agencies are mandated to deal with issues that affect the economic and societal well-being of their citizens; consequently, they have little incentive to concern themselves with extra-territorial matters, even when these matters can have a global impact.

The IOC is not above criticism in this context, having downgraded its Ocean Mapping Programme while introducing or continuing to promote initiatives that require accurate and reliable maps that describe the morphology and characteristics of the world seabed. Among others, these initiatives include: the prevention and reduction of the impacts of natural hazards; mitigation of the impacts of, and adaptation to, climate change and variability; safeguarding the health of ocean ecosystems; and sustainability of coastal and ocean environments and resources.

By providing adequate portrayals of the boundary conditions that constrain and influence processes that are the focus of IOC's present concerns, good bathymetric maps can improve mankind's understanding of the problems listed above while suggesting directions in which to seek solutions.

A few IBC projects have managed to maintain their levels of momentum through persistence and the determination to see things through to the end, benefiting from sponsorship support that was provided by visionary organizations which appreciated the benefits that would accrue to the international marine research community. It is to be hoped that the individuals and institutions involved in these undertakings are able to maintain their levels of commitment.

Conclusion

By dividing the world ocean into eight clearly defined project areas, the proposed restructuring of the IBC series could provide an effective global framework for systematic bathymetric compilations, thereby setting the stage for a well-ordered global ocean mapping programme. With the operational responsibility for each area assigned to an appointed Editorial Board that featured strong regional representation, it is to be expected that team members would be familiar with existing data sets within their compilation area, and that they would have access to that information for inclusion in a regional inventory of depth observations.

Moreover, team members should be in a position to maintain an awareness of current and upcoming survey missions within their project area, and to negotiate access to the resulting soundings in order to fill gaps in the regional data base. They should also be qualified to advise surveyors on optimum survey patterns for upgrading the state of bathymetric knowledge in the significant portions of the world ocean that remain woefully under-mapped. Such exchanges could usher in a new era of cooperation between the collectors and compilers of bathymetric information, and hasten the day when the global ocean floor has been thoroughly mapped.

For any of this to occur, however, the IOC will need to assume a strong position of leadership by revitalizing its Ocean Mapping Programme, and by urging member states to follow suit through involvement in, and sponsorship of, regional activities that will satisfy the information needs of mankind as it strives to better understand the marine realm of its home planet. Having demonstrated effectiveness in engaging its member states in other wide-ranging initiatives, the IOC clearly possesses the expertise and the moral authority to launch a global ocean mapping endeavour, beginning with the worldwide compilation of all available acoustic observations. It remains to be seen whether the IOC has the resolve.

9 Harmful algae: a natural phenomenon that became a societal problem

HENRIK ENEVOLDSEN

Henrik Oksfeldt Enevoldsen is a biologist who graduated from Aarhus University, Denmark. Now a staff member of the Secretariat of the Intergovernmental Oceanographic Commission of UNESCO, he has worked with international cooperation on harmful algae and capacity building for 20 years. He is a founding member of the International Society for the Study of Harmful Algae.

The last three decades have been marked by a new appreciation of the serious impacts of the marine phenomena we now call harmful algal blooms (HABs). These occurrences of toxic or harmful microalgae represent a significant and seemingly expanding threat to human health, fishery resources and marine ecosystems throughout the world. Many causes, both natural and anthropogenic, may be responsible for this dramatic expansion in HAB effects; it is likely that human activities in some cases are making the problems worse through increased nutrient inputs to coastal areas, transportation and discharge of ballast water, mariculture and physical modification (piers, dredging, ground water use of coastal waters) just to mention a few. Increasing impacts of HABs does not necessarily imply an increase in HAB occurrences; we just detect them more easily. The increased use of coastal waters (by increased human population, coastal settlements and economic activities in the coastal zone) unavoidably makes the impacts of HABs more evident.

In the mid 1980s the issue began appearing in the agenda of governing body meetings of international and regional organizations working within marine science, management, aquaculture and public health. In an example of how science responds to a societal need and how governments can collaborate to address a common hazard, the IOC, over the following years, developed programmes to address the situation.

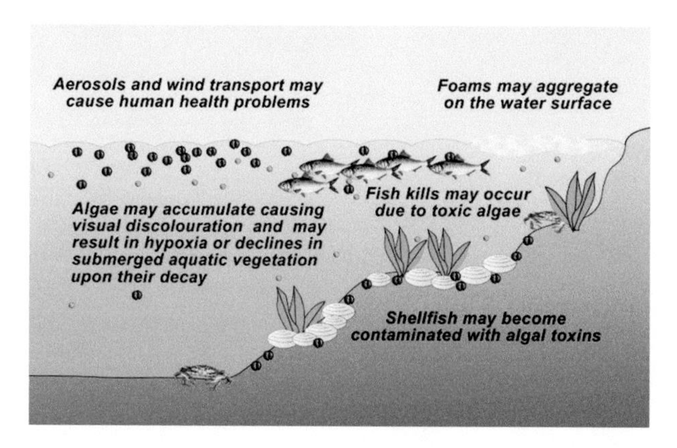

Figure 9.1 The impacts of HABs are numerous, and the effects may be felt by many components of the ecosystem. © GEOHAB.

The emergence of a problem

Phytoplankton blooms, microalgal blooms, toxic algae, red tides or harmful algae are all terms for naturally occurring phenomena. Many species of micro- and picoalgae are reported to form mass occurrences that discolour the water, while others increase in number to reach concentrations that don't discolour the water but are still dangerous or harmful. Many of these species are known to produce toxins. The scientific community refers to these events with a generic term, 'harmful algal bloom' (HAB), at the same time recognizing that, because a wide range of organisms is involved and some species have toxic effects at low densities, not all HABs are 'algal' and not all occur as 'blooms'.

Proliferations of microalgae in marine or brackish waters can cause massive fish kills, contaminate seafood with toxins and alter ecosystems in ways that humans perceive as harmful. A broad classification of HABs distinguishes two groups of organisms: the toxin producers, which can contaminate seafood or kill fish; and the high-biomass producers, which can cause anoxia and indiscriminate kills of marine life after reaching high concentrations. Some HABs have characteristics of both.

Although HABs occurred long before human activities began to transform coastal ecosystems, a survey of affected regions and of economic losses and human poisonings throughout the world demonstrates very well that there has been a dramatic increase in the impacts of HABs over the last few decades and that the HAB problem is now widespread and serious. When HABs contaminate or destroy coastal resources there are economic and health impacts and the livelihoods and sustenance of local residents are threatened.

The impact of harmful microalgae is particularly evident when marine food resources, e.g. aquaculture operations, are affected. Shellfish and in some cases finfish are often not visibly affected by the algae, but accumulate toxins in their organs. The toxins may subsequently be transmitted to humans and, through consumption

Figure 9.2 A flight division of the Swedish Coast Guard captured this image of a cargo vessel ploughing through an algal bloom near southern island of Landsort, near the outskirts of the Stockholm archipelago. Image by Swedish Coast Guard, 26 July 2008, reproduced with permission.

of contaminated seafood, become a serious health threat. Although the chemical nature of individual toxins is very different, toxins do not generally change or reduce significantly in amount upon cooking; neither do they generally influence the taste of the meat. Unfortunately, detection of contaminated seafood is not simple, and neither fishermen nor consumers can usually determine whether seafood products are safe for consumption. To reduce the risk of serious seafood poisoning, intensive monitoring of toxins is required for seafood products from harvested areas, as well as for the species composition of the phytoplankton.

In addition to posing serious health risks to consumers of seafood, some microalgae may have devastating effects on fish and other marine life, both wild and in aquaculture. Several species of microalgae can produce toxins which damage fish gills by haemolytic effects and result in extensive fish kills with major economic losses. A comprehensive economic analysis of the global impact of

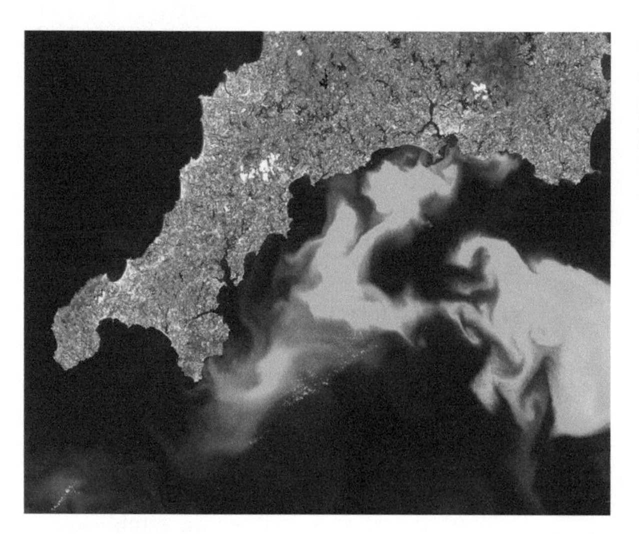

Figure 9.3 A NASA satellite image of a massive algal bloom producing cloud-like images in the ocean. Landsat image from 24 July 1999, courtesy of Steve Groom, Plymouth Marine Laboratory © NASA.

harmful algal events on the aquaculture industry is not yet available, but the economic losses from single events have on several occasions amounted to more than US$ 10 million. On one particular occasion a toxic bloom killed US$ 500 million worth of caged fish in Japan. Additional losses may occur from public disenchantment with seafood products due to misunderstanding and misinformation about harmful algal events.

In developing countries, seafood often constitutes an important source of food and protein, especially in coastal areas. With the increasing problems of overfishing, aquaculture may be needed as an alternative for the supply of seafood. However, it will be necessary to establish adequate surveillance programmes and quality control of the aquaculture products to minimize the risk of seafood poisonings and economic losses. These countries will often require expert assistance from countries which have longstanding experience in this matter. There is currently no international record of the number of incidents of human intoxication caused by contaminated seafood. The numbers available are undoubtedly underestimates, as many cases and even fatalities can be assumed to pass undiagnosed and hence unreported officially.

A number of human syndromes are presently recognized to be caused by consumption of contaminated seafood: amnesic shellfish poisoning (ASP), ciguatera fish poisoning (CFP), diarrhetic shellfish poisoning (DSP), neurotoxic shellfish poisoning (NSP), paralytic shellfish poisoning (PSP) and azaspiracid shellfish poisoning (AZP). Other threats to human health are posed by blue-green algal toxins in drinking water which may cause severe liver damage or be tumour promoters, and by toxic aerosol and contact problems caused by benthic marine dinoflagellates.

Figure 9.4 Shellfish are particularly vulnerable to the toxins produced by HAB events. Here women in South Africa are pictured gathering brown mussels. © Kathleen Reaugh/Marine Photobank.

A need for international cooperation

In the years before the mid 1980s the knowledge on harmful algae was primarily held by people working with traditional algal taxonomy, toxin chemistry and phytoplankton ecology. There was neither an established research community nor any extensive multidisciplinary cooperation to address harmful algal events, and possible mitigation of their impacts. The expertise was scattered both nationally and globally and particular expertise in the causative organisms was lost or about to be lost as focus on taxonomy had been under severe pressure in favour of other newly emerging biological disciplines. Similarly, only a few countries had well-established monitoring and management programmes and few governmental or non-governmental organizations had the issue on their agenda.

The earliest initiative at a regional level on HABs was by ICES, and began in response to outbreaks in Danish and Norwegian waters in 1966. The Danish State Biologist Chr. Vagn Hansen was prominent in this initiative, and, many years later, became a key figure in putting HABs on the agenda of the IOC and in ensuring long-term Danish funding for an international programme. The 1966 episode was soon followed by major PSP events in the UK (north-western North Sea) in 1968, and the USA (western Gulf of Maine) in 1972. The latter provoked the First International Conference on Toxic Dinoflagellate Blooms, convened in Boston, Massachusetts, in November 1974. The discovery and description of the involved toxins and development of toxicity testing is a whole chapter by itself with pioneers such as Hermann Sommer (Switzerland/USA) and Karl F. Meyer (Switzerland/USA) who developed the mouse bioassay protocol, which has played a central role in subsequent studies and monitoring activities to protect public health, and Edward J. Schantz (USA) and his colleagues who were the first to extract and purify saxitoxin, the main toxin causing paralytic shellfish poisoning.

A milestone in the development of international cooperation within the growing HAB research community was the International Symposium on Red Tides held in 1987 in Takamatsu, Japan, under the initiative and leadership of Tomotoshi Okaichi and Takahisa Nemoto of Japan. The Conference marked the first broad support and co-sponsorship of several governmental and non-governmental organizations, reflecting recognition of the issue at many levels of society. The Symposium was followed by an IOC workshop on international cooperation in the study of 'red tides and ocean blooms'. This was based on recognition by the IOC Assembly earlier in 1987 that it had the appropriate mandate and position to take the lead in developing an overview of relevant research, to formulate research strategies for a global programme on harmful algae and to identify major research topics.

By the late 1980s the scientific community had grown significantly compared to the 1960s and 1970s, and had attracted both scientists from other fields and young scientists.

Early in 1990 an IOC-FAO/OSLR ad hoc group of experts on harmful algal blooms was formed and a group of key scientists were invited to Paris to draft a programme based on the recommendations of the Takamatsu workshop. Many of the people involved were, or became, leaders in the field as well as in IOC activities. This was also the time when the term 'harmful algal bloom' was officially chosen as the descriptor for the field, in preference to 'red tide'.

The draft plan was endorsed by the IOC and was fully elaborated at an IOC-SCOR Workshop following the Fifth International Conference on Toxic Marine Phytoplankton in Newport, Rhode Island, USA, in November 1991. The workshop delivered the final focus for the IOC efforts and formulated the overall goal of the HAB programme: '*To foster and organize the management and scientific research of Harmful Algae Blooms in order to understand the causes, predict the occurrences, and mitigate the effects*'. The special directions were to develop support for a decentralized secretariat for the programme in a research institution; initiate the publishing of a newsletter; and publish a manual on harmful algae and an on-going directory of experts for harmful algae.

A major requirement identified by institutions and governments was to develop a comprehensive programme component offering training and capacity enhancement. To develop this component and to identify priority issues, the IOC and the Bremen Maritime Training Centre in Germany co-sponsored an international workshop on capacity development activities.

A characteristic of the IOC activities on harmful algae has from the early days been close cooperation and interaction with other governmental and non-governmental organizations, and the programme was developed as a joint undertaking with FAO, with the assistance of SCOR. Close cooperation with ICES was seen as essential for

Figure 9.5 A fish kill on the West Coast of South Africa caused by a high biomass bloom of the dinoflagellates *Prorocentrum micans* and *Ceratium furca* (P. Glibert and G. Pitcher, *Oceanography* **18** (2), 2005.

reaching out and involving the scientific community. Therefore the strengths of several governmental and non-governmental organizations were combined in the design of an international activity to deliver specific and concrete results through establishing or enhancing HAB research, monitoring and management. Thomas Osborn, from the Johns Hopkins University (USA), was seconded to the IOC at this time and his strategic vision and ability were an important factor in the first years of the programme. From the outset, he recognized and focused on the critical importance of strengthening the collaboration and interface between the physical oceanography and biological, organism- and species-related research communities.

Raised expectations and challenges

The emergence of an international programme plan in 1992 raised expectations of what was to be delivered, and funding remained a key issue in the ability to deliver tangible results. To govern the new programme, and to identify the resources required, the IOC and FAO had established an Intergovernmental Panel on Harmful Algal Blooms (IPHAB), which met for the first time in June 1992 with the participation of 14 nations. Today, 43 nations and organizations are members of IPHAB.

Although the need for an intergovernmental mechanism for the direct governance and funding of a programme, to address harmful algae and their impacts, was recognized at the level of scientists and managers, it was not necessarily reflected in national infrastructures. This posed a challenge in finding a homogenous composition of people for IPHAB at the appropriate experience and level to be effective. The scientific community working on the issue was, and still is, very multidisciplinary, and can be found in a broad variety of research institutions under marine, food, fisheries, public health or agriculture government agencies. Furthermore, rather than dealing with international cooperation in ocean science, these agencies are more oriented towards the programmes of the UNEP, WHO, FAO or IMO. Considering this diversity, it is striking that the IPHAB, with a composition of delegates ranging from academia to governmental managers, has managed to fulfil its expected role so successfully. The existence of an intergovernmental mechanism for HAB has raised the visibility and awareness of the issue at the national and global levels.

Several of the most active Member States, in terms of activities and funding, are represented on IPHAB by scientists who have managed to use their position on IPHAB to achieve commitment in higher levels of their governments. In particular, this includes the provision of funds supporting the IOC Science and Communication Centres on Harmful Algae at the University of Copenhagen (Denmark) and at the Spanish Institute of Oceanography in Vigo (Spain). These Centres have proven to be efficient and visible nodes of expertise in the development and coordination of international collaboration and networking on harmful algae.

As is the case for most multilateral mechanisms, IPHAB has difficulty in generating adequate funds for the necessary international collaboration on harmful algae and for the provision of improved tools for management and mitigation. The Member States that established the multilateral programme to address the recognized priority issues face difficulties in providing the necessary funds for its implementation. The effort required to achieve financial contributions cannot always be generated from the limited human resources available within the IOC Secretariat. Sporadic and short-term commitments of funds do not allow for continuity and the pursuit of a clear strategy. Furthermore, longer term commitments enable a platform to be developed where specific contributions can be accommodated in a meaningful way. For example, the IOC HAB Programme would not have been able to maintain a coherent framework for its activities without the stable support of Danida (Denmark) from 1991 to 2007.

Figure 9.6 Agriculture runoff and wastes from other human activities are carried by rivers into the sea, where excessive nutrients in the coastal zones can contribute to algal blooms. © Steve Shoup/Shutterstock.com.

Where the nature of IPHAB and its membership has been weakest is in relation to linking effectively with the other major specialized UN agencies. FAO was an initial sponsor but withdrew in 1993 and only in 2009 showed renewed interest in collaboration via IPHAB. Collaboration and involvement in the issue with WHO has also been lacking as that Agency has the issue embedded in programmes devoted to drinking water, safe use of recreational waters and desalination. This limited interaction is regrettable and a reflection of a broader problem where the structures of the different agencies may not match well although they address the same, or different, aspects of the issue. However, collaboration with other UN agencies and governmental organizations has been more direct and fruitful. In particular with the International Atomic Energy Agency (IAEA), UNEP (Regional Seas), the North Pacific Marine Science Organization (PICES) and, as already mentioned, ICES.

The IPHAB Chairs since 1992 (Bernt Dybern (Sweden), Adriana Zingone (Italy), Beatriz Reguera (Spain) and Leonardo Guzman (Chile)) have therefore been challenged to match the stated priorities and needs of its members with limited resources and funding.

Concerted actions and enhanced capacities

An intergovernmental approach to an issue requires coordinated, concerted and complementary action by Member States. An initial step in focusing international research on HAB was a SCOR-IOC co-sponsored Working Group on the Physiological Ecology of Harmful Algal Blooms that led to a NATO funded Advanced Study Institute in Bermuda in 1996. The coordinated, concerted and complementary

Figure 9.7 The greatest volume of waste discharged into the ocean is sewage. This outfall discharges 15–20 million gallons of sewage upstream of a coral reef. © Steve Spring/Marine Photobank.

action was further applied in the design of an international HAB research programme. The goal was to improve the prediction of HABs by determining the ecological and oceanographic mechanisms underlying the population dynamics of harmful algae, integrating the related biological, chemical and physical studies, and providing adequate observation and modelling systems. The IOC-SCOR international research programme on the Global Ecology and Oceanography of Harmful Algal Blooms (GEOHAB) established in the late 1990s has subsequently developed several core research projects on HABs in different environments, such as upwelling systems, eutrophic systems, stratified systems, etc. The key characteristic of GEOHAB is that it compares similar systems or similar species, recognizing there is no generic approach to model or describe HAB ecology, and that development of forecasting abilities must rely on an understanding of the species in a given site-specific system. GEOHAB aims at delivering improved observation and forecasting abilities which will lead to routine HAB observing systems.

Unlike the physical oceanographers or other members of the chemical and biological oceanographic disciplines, the HAB scientific community does not have a track record of larger international research programmes, large multi-investigator research projects or a tradition for strongly standardized methodologies. Oceanographic research programmes have their roots in disciplines where research focus and methods are probably more uniform and fewer than in multidisciplinary HAB research. Furthermore many HAB researchers operate alone or in small teams and, as mentioned earlier in this chapter, are scattered throughout very different institutions. An exception is Don Anderson's research group at the Woods Hole Oceanographic Institution (USA), whose long-standing research programme in the Gulf of Maine is one example of how far our understanding of a given system can develop and how far modelling and prediction of

HABs can be taken. This group in particular nourished and supported the ideas and concepts that led to GEOHAB, which in turn is developing and gaining involvement in the research community. In recognition of his work, Dr Anderson was invited to present the Anton Bruun Memorial Lecture at the IOC Assembly in 2005. Patrick Gentien from IFREMER (France) is another scientist who strongly influenced the concepts of GEOHAB with his early focus on the necessity of a species-specific approach and recognition of the diversity of factors controlling population dynamics in different systems.

The GEOHAB results published as individual papers, special journal issues and reports are contributing to a higher level of understanding and refinement of efforts to deliver science that is applicable to management, reflecting both advances in science and in the application of results. Nevertheless, the recognition of the benefits of HAB research to governmental and social programmes remains a challenge in terms of the needed investment in both national and multilateral activities.

Another concrete outcome of intergovernmental cooperation over the past 15–20 years is the provision of training opportunities to upgrade the competence of personnel, for both research and regulatory monitoring of HAB events. A close partnership among organizations and national institutions with recognized international expertise has led to publication of manuals on methodology and numerous training opportunities encompassing a broad variety of regions, levels and purposes. Gustaaf Hallegraeff (Australia), Don Anderson (USA) and Allan Cembella (Canada) delivered a milestone by editing the first comprehensive manual on research methodologies for harmful algal research, which was published by UNESCO. Prominent amongst the institutions which pioneered in offering international capacity building are the University of Copenhagen (Denmark) with Jacob Larsen and Øjvind Moestrup, University of Tokyo (Japan) with Yasuwo Fukuyo and the Spanish Institute of Oceanography in Vigo with Beatriz Reguera and Santiago Fraga. Together these institutions have trained more than 600 individuals. In addition, Dr Fukuyo was already a pioneer in capacity building at the regional level in the 1980s, as he commenced systematic training in HAB observation and species identification both in bilateral projects and within the IOC Sub-Commission for the Western Pacific (WESTPAC). Furthermore there has been a close and important cooperation with training programmes with relevant regional and global organizations, including IAEA, FAO, PICES and ROPME, and these have led to a global network of institutions, organizations, trainers and trainees that would otherwise not have existed. A key component in building capacity globally is communication. An important tool is the IOC newsletter 'Harmful Algae News', which has been published since 1992 thanks to a remarkable commitment by the editor Timothy Wyatt.

Figure 9.8 Harmful algal blooms sometimes colour the water red and are known as 'red tides'. Here a red tide hits the coast near Seabeck on Hood Canal, Washington State, USA. © Don Paulson.

The value of open access to global datasets on the occurrence of harmful algal events and the biogeography of potentially harmful microalgae is obvious to scientists, managers and policy makers. Such data have broad interest for research purposes, for risk assessment in planning of mariculture, ballast water issues, effects of climate change, tourism development, etc. However, obtaining the relevant data from national environmental and food safety monitoring programmes remains a major problem. Overcoming this apparent paradox is clearly a task for intergovernmental cooperation and one that has received priority from the outset. The work began in close cooperation with the International Council for the Exploration of the Sea (ICES) in the early 1990s and later expanded through the North Pacific Marine Science Organization (PICES) and IOC regional HAB networks. The HAB data compilation is developing as a multi-partner undertaking within the framework of the IOC International Ocean Data Exchange (IODE) and is furthermore linked with the World Register of Marine Species and OBIS, the latter now part of the IOC.

It is difficult to quantify how the intergovernmental activity on HABs has improved the capacity to mitigate the effects of harmful algae at the national level, but clearly, the development and identity of the HAB research and managerial community has been strongly influenced by, and integrated with, the activities of IOC and its partner organizations. Certainly, the provision of an international newsletter, regional networks, various training courses and the establishment of an international NGO (International Society for the Study of Harmful Algae, ISSHA) constitutes a visible global framework for the HAB research and management community.

Challenges ahead

Both scientific and organizational challenges will continue into the future. The development of low cost *in situ* automated and high-resolution operational observing systems for HABs is a long-term goal. Such systems will provide the on-going data sets necessary to document trends and deliver forecasts. HAB observations should eventually be integrated with other operational observing systems for coastal areas. In relation to securing food safety, the recurring emergence of new toxins or toxin analogues continues to challenge routine monitoring.

In terms of research, our knowledge of how climate change will affect the biogeography of potentially harmful species is limited, as is our understanding of the ecology and population dynamics of benthic dinoflagellates, despite the fact that ciguatera is the most severe of the HAB poisoning syndromes both in terms of public health and impact on local economies. A key issue is why some species of microalgae are toxic in one geographical location or sub-population and not in another. We also face new unexpected problems with toxic aerosols produced by other benthic dinoflagellates in warm temperate areas, such as the Mediterranean Sea.

On the plus side, the critical mass of scientists, the general level of knowledge and understanding, and the technologies available have dramatically increased over the past 15 years and thus a good basis of scientific knowledge exists to devise and implement observation and management/mitigation methodologies.

The main organizational challenge is to continue to involve the scientific community as well as managers and governments, so that all three communities perceive the activities and services delivered by the intergovernmental mechanism as relevant and timely. The IOC and partners must continue to provide the intergovernmental cooperation required to deliver the needed research, to translate scientific results for management and to facilitate and deliver enhanced capacity at national institutions. The success of these programmes is essential to win the trust and support of funding agencies in competition with other marine science priorities related to resource management, integrated observing systems, food safety, etc. This implies that many aspects of the HAB issue are best addressed by being fully integrated with these related programmes, gaining significant added value through the development of cross-disciplinary and cross-sectorial linkages combining science, management practices and policy. This presents a continuous challenge for the IOC and its IPHAB.

The IOC is predominantly driven by the interest of national agencies. The delegates attending the governing bodies of the IOC may not have HABs or even other marine biology issues within their respective mandates. In times of publicly visible priorities like climate change, tsunami warning systems and emerging major undertakings like the UN regular process for the assessment of the marine environment, it is incumbent on the HAB research and management community to keep a visible place on the intergovernmental agenda. The international HAB community can make the case that mitigating the effects of harmful algae is a strong and well-justified driver for intergovernmental cooperation, and that this area provides a close match between societal needs and science.

10 Non-governmental international marine science organizations

ELIZABETH GROSS

Elizabeth Gross trained as a biological oceanographer. She was Executive Director of the Scientific Committee on Oceanic Research (SCOR) from 1980 to 2000. Since her 'retirement' she continues as part-time Finance Officer for SCOR, and takes on special projects, such as conference organization.

Non-governmental marine science has an expression in the work of the Scientific Committee on Ocean Research (SCOR). While they are very different organizations in many ways, the Intergovernmental Oceanographic Commission and the Scientific Committee on Oceanic Research (SCOR) have had close ties for all of the 50 years that both of them have been in existence. Their origins are closely linked and many of the same people (Roger Revelle, George Deacon and Warren Wooster, among others) were involved in the creation and early years of both organizations.

Born out of the International Geophysical Year, SCOR was established in 1957 by the International Council of Scientific Unions (ICSU, now known as the International Council for Science) 'to further international scientific activity in all branches of oceanic research' (SCOR Constitution). SCOR was the first of the ICSU interdisciplinary bodies to be established, and was the first truly global non-governmental organization in ocean science. The IOC was created just 3 years later.

In 1960, UNESCO organized an Intergovernmental Conference on Oceanographic Research, held in Copenhagen, which recommended the establishment of an Intergovernmental Oceanographic Commission and defined the scientific advisory relationship of SCOR to UNESCO and the IOC. It proposed four main goals for the Commission: the last of these was 'to facilitate the performance of international programmes (e.g. the International Indian Ocean Expedition)'.

Figure 10.1 Tube worms feeding at the base of a black smoker chimney hydrothermal vent. © Image courtesy of NEPTUNE Canada/Tunnicliffe/Juniper.

Recognition that the scientific problems of the ocean require a truly interdisciplinary approach was embodied in ICSU's plans for the International Geophysical Year (IGY) of 1957–1958. Accordingly, SCOR's first major effort was to plan a coordinated international approach to the least-studied ocean basin of all, the Indian Ocean. This planning process was the major focus of SCOR during the first few years of its existence. While it could not be achieved within the time frame of the IGY, the International Indian Ocean Expedition (IIOE) of the early 1960s was certainly a product of that effort. The need for intergovernmental coordination, communication, training and management of the project rapidly became obvious. In the first example of significant cooperation between SCOR and IOC, these functions were transferred to the Commission, while SCOR retained scientific oversight of the project (http://www.scor-int.org/history.htm).

As well as planning the IIOE, from its beginning SCOR set up Working Groups to examine specific questions in ocean science, methodology, data management, etc. From the start, the small, nimble and relatively short-lived SCOR Working Groups (WGs) have dealt with narrowly defined scientific objectives. These groups have always been set up in cooperation with other international organizations when appropriate; often with IOC. As recently noted by former SCOR Executive Secretary George Hemmen, this concept of WGs with limited tasks and limited life was to become the life-blood of SCOR and still prevails today.

For many years, UNESCO, and later the IOC, provided a venue for the publications of these groups through various UNESCO series, such as the *Technical Papers in Marine Science*. In the last two decades, however, most SCOR WGs have published their scientific findings in the primary scientific literature, as review articles or special issues of journals.

The work of SCOR over the years covers too broad an area to be dealt with in this contribution, but some highlights of achievements with lasting impacts on the field of oceanography in general and on the IOC programmes in particular can be mentioned.

Several working groups, for example WG 67 on Oceanography, Marine Ecology and Living Resources, fed directly into the development of IOC's scientific programmes. Its final report, presented to an IOC Assembly in 1982, provided an early scientific framework for the programme on Ocean Science and Living Resources (OSLR). Another, WG 97 on the Physiological Ecology of Harmful Algal Blooms, provided the stimulus for the planning and implementation of the joint SCOR and IOC scientific programme on the Global Ecology and Oceanography of Harmful Algal Blooms (GEOHAB).

In the arena of physical oceanography and climate, two WGs in the early 1980s were very closely tied to the international research programme known as TOGA, for Tropical Ocean and Global Atmosphere. These were WG 55 on the Prediction of El Niño and WG 56 on Equatorial Upwelling Processes. SCOR and IOC worked closely together on topics such as these.

In the coastal marine science sphere, there was also close collaboration between SCOR and IOC for WG 112 on the Magnitude of Submarine Groundwater Discharge and its Influence on Coastal Oceanographic Processes, for which IOC co-sponsored several important field experiments and published the report 'Submarine Groundwater Discharge: Management Implications, Measurements and Effects' in the *IOC Manuals and Guides* series.

These are but a few of the many examples of the small Working Groups fostered by SCOR providing scientific expertise to support the intergovernmental activities of IOC.

But a huge change was on the horizon for SCOR, the IOC and for oceanography in general in the early 1980s with the gradual emergence of truly global programmes. With advances in computing power and modelling skills, it had become possible to construct global models of the ocean, crude as they might appear to us today. In an age when there were often high walls between the disciplines, physical oceanographers took offence when the atmospheric scientists made the entire ocean a 'boundary condition' in their global models. Likewise, the biologists were upset when the physical oceanographers put all oceanic biology into a single 'black box' in their global models.

Even a mathematically challenged person could see from the graphics presented at many scientific meetings held throughout the 1980s that models were getting more complex and more interdisciplinary. Coupled ocean–atmosphere circulation models were developed, then models in which the physics and biology in the oceans were coupled, and so on. Increased computing power made more and more complex models possible for predictive purposes and other scientific applications.

The advent of satellite remote sensing of the marine environment occurred at the same time. In 1978, two satellites were launched that would revolutionize our view of the oceans and the collection of oceanic data. They were SEASAT, which produced the first reasonably accurate sea surface altimetry data, even if it failed after only 105 days, and NIMBUS-7, with the Coastal Zone Colour Scanner (CZCS), which gave us the first global views of ocean biology.

It has been said that satellite instruments can collect in 1 minute the data that would have taken a single oceanographic vessel 10 years to collect. While research ships are still necessary for process studies and for deep ocean research, there can be no doubt that remote sensing has changed oceanography forever.

These two developments made possible the emergence of truly global oceanographic research programmes. While the meteorologists already had a worldwide data collection network to support weather forecasting and, by about 1980, they had successfully completed the First GARP Global Experiment (FGGE) under the auspices of the Global Atmospheric Research Programme (GARP), nothing comparable had been attempted in the oceanographic community. However, because of the costs of operating deep sea research vessels, marine scientists were accustomed to cooperating and coordinating efforts in order to maximize the efficiency with which they used these expensive facilities.

In the late 1970s, there was growing concern over the likelihood of global climate changes due to the increasing levels of carbon dioxide and other greenhouse gases in the atmosphere as a result of human activities. Roger Revelle, who had been so instrumental in the establishment of both SCOR and the IOC, was a leading proponent of the need for truly global research programmes to elucidate these concerns. Accordingly, the World Meteorological Organization and ICSU transformed GARP into a World Climate Research Programme (WCRP).

SCOR was well placed to contribute to this expanded effort having, a year or so earlier, created with IOC a Committee on Climate Changes and the Ocean (CCCO). Planning now started for major ocean science studies within WCRP notably the Tropical Ocean and Global Atmosphere (TOGA) programme and the World Ocean Circulation Experiment (WOCE).

TOGA was focused on the links between the equatorial Pacific Ocean and the global atmosphere and their impacts on the phenomenon known as El Niño. As a result of TOGA, our ability to predict El Niño has improved dramatically, and an observing system to collect data to be assimilated into these predictive models has been put in place.

WOCE, on the other hand, had as its ultimate goal a better understanding of the circulation of the ocean, from the surface to the deep water. This would underpin the development of models for the prediction of longer term climate change and the

Figure 10.2 Satellite instruments can measure the height of the sea surface to such accuracy that we can observe the thermal expansion of sea water as it warms and cools. These remote sensing measurements advanced the study of large-scale phenomena such as El Niño, seen in the upper image in 1997, and its converse, La Niña, shown in 1999. White, red and orange colours denote higher sea surface or warmer waters. Blue and purple shades represent cooler water. © NASA.

establishment of a system for the collection of the data necessary to test them. As a result of WOCE, we have a much better understanding of the areas of the ocean most sensitive to climate change, such as the regions of deep water formation in the sub-arctic North Atlantic.

These two important international, global-scale research programmes have provided governments with the data for models to forecast the physical responses of

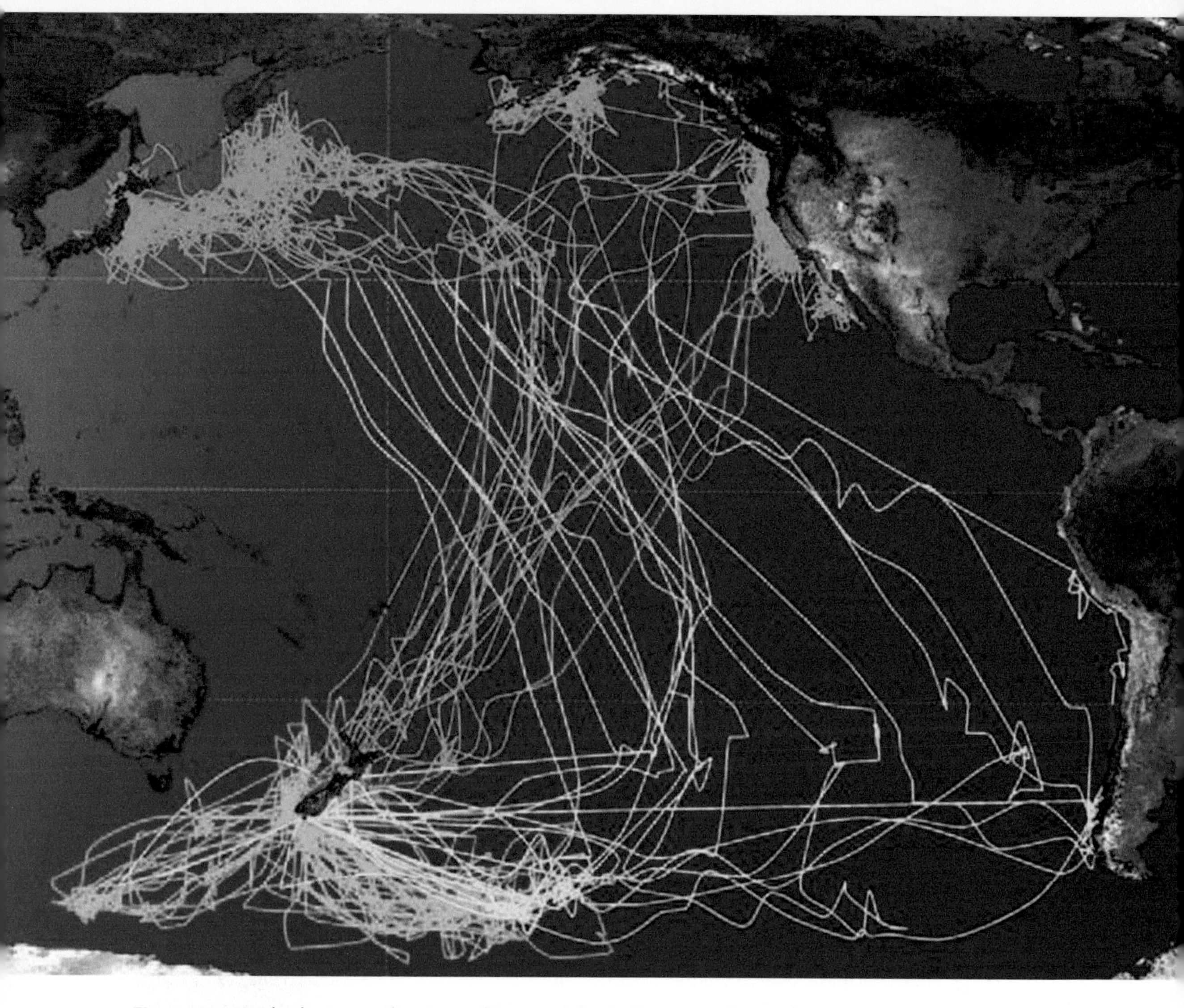

Figure 10.3 Seabirds are another type of top predator in the ocean. Sooty shearwaters perform the longest migrations recorded electronically in the animal kingdom, preying on small fish in the most productive areas of both the North and South Pacific basins. These birds are in decline and may serve as a useful indicator of oceanic health. © National Academy of Sciences, USA (2006).

the ocean to global warming, and the information necessary to put in place operational ocean observing systems.

By the mid-1980s, it was clear that more than physical understanding of the ocean was needed to understand its complex responses to a rapidly changing Earth. How were the complex chemistry and biology of the ocean involved in the absorption of anthropogenic carbon dioxide from the atmosphere? Did the ocean have a limitless capability to absorb this excess CO_2? How would marine biogeochemistry change in a greenhouse environment?

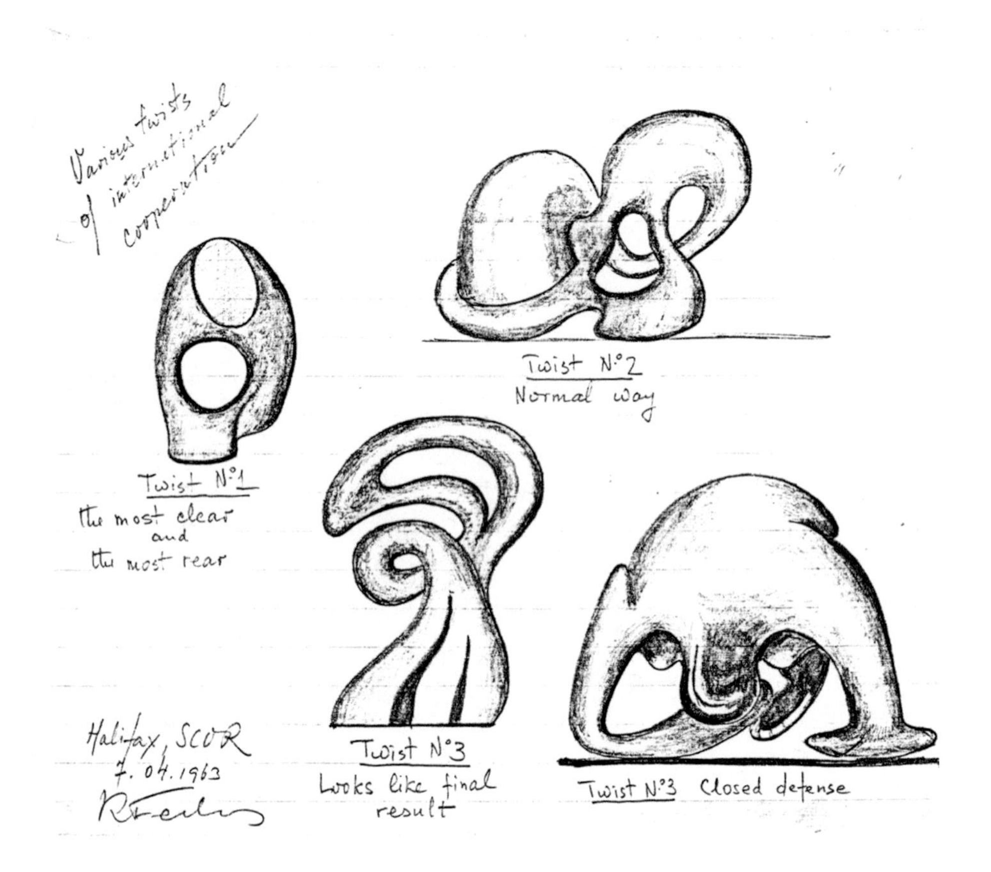

Figure 10.4 The path of international scientific cooperation is not always an easy one. These thoughts on the topic were committed to paper by Konstantin Fedorov (1927–1988) during a 1963 SCOR meeting in Halifax. Fedorov served both as Secretary of IOC from 1963 to 1970, and as President of SCOR from 1976 to 1980.

At the same time, oceanographic research was becoming increasingly interdisciplinary. The advent of 'ocean colour' satellite instruments made it possible to obtain a global synoptic view of the biological productivity of the surface ocean and large-scale phenomena could be observed for the first time, such as the biological response of the ocean to El Niño, which was being studied so intensively in the TOGA programme, and the intense interannual biological variability of the north-west Indian Ocean associated with monsoonal overturning. What is the role of the phytoplankton, their uptake of carbon dioxide, subsequent predation by zooplankton, recycling and eventual burial of carbon in marine sediments in the context of climate change?

POGO

In 1999 the Directors of three major ocean research laboratories, Scripps, Woods Hole and Southampton, met at the IOC in Paris under the leadership of Charles Kennel, Director of Scripps. They established a Partnership for Observation of the Global Oceans (POGO), which ten years later has a membership of more than 30 ocean research institutions worldwide. POGO Directors meet regularly to plan long-term cooperation in global ocean research. By hosting students in member laboratories, POGO is an important contributor to capacity development, for marine science and its applications.

National programmes were being developed to answer questions like these. SCOR was ideally placed to respond rapidly to the need for international coordination, development of standard protocols for measurements and the development of collaborative process studies in widely dispersed regions of the oceans. So, in 1988, was born the Joint Global Ocean Flux Study, with the goal of understanding the vertical fluxes of carbon and other biogenic elements in the ocean, and to predict the response of oceanic biogeochemical processes to anthropogenic perturbations, in particular those related to climate change.

Not long after it began, JGOFS was adopted by the International Geosphere-Biosphere Programmes (IGBP) as one of its Core Projects of global change research.

JGOFS developed rapidly and, within 18 months of the first international planning meeting, research vessels were conducting the first international cruises. Carefully selected regions of the ocean were intensely studied and a global survey of carbon dioxide was conducted in collaboration with WOCE. The programme was completed in just over a decade. While JGOFS did not answer every question about the fluxes of carbon in the ocean, it was an enormous success, and it left a legacy of well-cared-for data sets, coupled models and the groundwork for future international and truly interdisciplinary global-scale research programmes dealing with the ocean and climate change.

The programmes that followed JGOFS all have different foci. The programme on Global Ocean Ecosystems Dynamics (GLOBEC) was co-sponsored by SCOR, IGBP and IOC from its inception. The aim of GLOBEC is to advance our understanding of the structure and functioning of the global ocean ecosystem, its major subsystems and its response to physical forcing so that a capability can be developed to forecast the responses of the marine ecosystem to global change (http://www.globec.org).

Figure 10.5 A scientist studies specimens in a lighted aquarium on Lizard Island Reef, Australia. Photo by Gary Cranitch, Queensland Museum; image courtesy of Census of Marine Life, http://www.coml.org

Other SCOR programmes, such as the Surface Ocean Lower Atmosphere Study (SOLAS) and Integrated Marine Biogeochemistry and Marine Research (IMBER), are also part of IGBP. The newest SCOR programme is GEOTRACES, which has as its goal to identify processes and quantify fluxes that control the distributions of key trace elements and isotopes in the ocean, and to establish the sensitivity of these distributions to changing environmental conditions.

The intention here is not to describe this suite of newer programmes in detail, but to make the point that fundamental scientific understanding must form the basis for the design of all components of the operational ocean observing system that is discussed elsewhere in this volume. Research efforts also underpin the development of predictive models for ocean forecasting relating to climate, marine fisheries, HABs and so on.

The communities served by the IOC and SCOR are very different, but their interests frequently converge.

On the one hand, IOC is made up of its Member States, whose representatives determine and implement policies and decisions that are binding on their governments. By their nature, intergovernmental organizations can seem to be deliberate and somewhat inflexible.

Non-governmental organizations like SCOR, on the other hand, represent the research community. Not having the force of governments behind their decisions, they can rely only on the good will of members to fulfil their commitments to contribute to international research efforts.

The flexibility and ability to respond quickly to new scientific issues is the strength of the NGOs like SCOR. And these characteristics are what attract scientists to them – to commit their time and effort to working with others towards common objectives. But, the strength of the IGOs lies in their power to implement recommendations arising from the scientific results and to develop policies based on the best available scientific information.

While the partnerships between these two very different types of organizations are not always easy ones, they are critical if society is to benefit from the best possible scientific information.

Census of marine life

In the late 1990s several leading marine scientists shared their concerns with the Alfred P. Sloan Foundation, a private USA philanthropy, that humanity's understanding of what lives in the oceans lagged far behind our desire and need to know. Some emphasized the chance for exciting discoveries about the diversity of life in the oceans. For example, ichthyologists had identified about 15,000 species of marine fishes but also believed about 5,000 species remained to be discovered. Other researchers highlighted the importance of establishing baseline information on the distribution of marine life. For most of the 240,000 or so known marine animals, we lacked reliable maps of their range, crucial for designation of marine protected areas. Still other researchers pointed to the changing abundance of economically valuable species and the need for improved management of fisheries. They noted increasing exploitation of largely unsurveyed areas such as the continental margins and sea mounts as well as violent debates about numbers of supposedly well-known species such as cod, tuna and salmon.

Essentially, marine biology and especially studies of marine biodiversity needed to join fields such as high-energy physics and human genetics in carrying out 'Big Science' that would integrate thousands of researchers concerned with all taxonomic groups from microbes to mammals and all ocean realms from near-shore to mid-ocean and the abyssal depths to the surface. The community converged on a strategy to address its concerns: conduct a worldwide, decade-long Census to assess and explain the diversity, distribution and abundance of marine life. Clearly, such a programme would ultimately depend on funding by national government agencies and the operational framework of international intergovernmental organizations, such as the Intergovernmental Oceanographic Commission (IOC) and Global Biodiversity Information Facility (GBIF), and international non-governmental organizations such as the Scientific Committee on Oceanic Research (SCOR) and Scientific Committee on Antarctic Research (SCAR). Sloan agreed to anchor the programme financially on the condition that the

Continued…

. . . Continued

scientific community successfully engage these national and international groups.

The leading mechanism to engage national agencies became 'National and Regional Implementation Committees' closely associated with the International Scientific Steering Committee that planned and monitored the Census as a whole and included top officers of the IOC, FAO and other key partners. Engagement with the international governmental and nongovernmental organizations developed around fulfillment of specific functions required to accomplish the Census, for example, data management with the IOC and technology diffusion with SCOR. The Census always saw itself as a programme that could be implemented within the existing organizational ecology of ocean sciences, and indeed could contribute to its strengthening, for example, by developing biological elements of the Global Ocean Observing System. This has proven to be the case.

Jesse H. Ausubel
Vice President, Alfred P. Sloan Foundation

Part IV Observations and Data: Preface

Scientific advances over the past 50 years have given us an understanding of the physical, chemical and biological processes that govern ocean behaviour. Our understanding is still partial and imperfect in many aspects, but nevertheless, good enough to enable realistic modelling and forecasting of future changes. Perhaps the major scientific challenge for the next 50 years is to improve the accuracy and availability of these forecasts. The challenge for governments is to set in place procedures to act on these forecasts and warnings to mitigate their impacts.

The basis for reliable modelling and forecasting is a reliable set of appropriate observations, first as initial conditions for the models, and then to check and adjust their projections as they develop. Good observations cost money, whether for satellite sensing, ship measurements or buoy technology. While scientific funds paid for the experimental work in several of the major coordinated programmes (IIOE, TOGA, WOCE, Global Ocean Data Assimilation Experiment (GODAE)) permanent operational observations are more appropriately funded by operational agencies.

The issue of funding observations systems is ongoing. There is no established customer base to use for market research and planning, though many of the forecasting products undoubtedly have commercial value. Establishing markets for completely new applications is difficult, and market research can only take us so far. As Henry Ford said of his newly developed motorcar: if I had done market research, people would have asked for a faster horse! We need governments to act as proxy customers for ocean forecasting. The next chapter describes how progress is being made for a truly Global Ocean Observing system.

Just as for observations, the management of data has become a major oceanographic activity. Modern technology produces quantities of data unimaginable 50 years ago; fortunately it has also given us the ability to analyze and understand the data, and to store it for possible future use. Reliable marine measurements increase in value as time passes, especially where we are looking for trends. For example, sea level data from the nineteenth century is now throwing new light on mean sea level changes over the past 100 years.

To maximize the use and value of data requires it to be used as many times and in as many ways as possible. Ideally the nations and organizations that collect marine data would, after using it for their own purposes, make it freely available for others to use. Not unexpectedly, they may feel entitled to some contribution towards the cost of the original measurement. But collecting money like this seldom works: instead people, including scientists, tend to make do without… which actually minimizes the value extracted from the data.

There are some lessons to be learned for ocean data exchange from meteorology and the data policy of the World Meteorological Organization. A clear distinction is made between the basic data and the products, which may be prepared from the data. The latter are considered valued added, and there may be a charge. These are still early days in the development of effective data exchange processes. Nevertheless, over the past 50 years, data management has gone from lists of numbers on paper to gigabytes of digitally stored and accessible information. National marine data centres are more and more common. Their cooperation through the International Oceanographic Data and Information Exchange (IODE) programme of the IOC is discussed in the second chapter in this section.

Both observations and data management are essential components in the operation of any future ocean observing and forecasting systems. As in many other aspects of ocean cooperation the technical possibilities are at present well ahead of the management structures we have in place to maximize the potential benefits.

11 Ocean observations: the Global Ocean Observing System

PETER DEXTER AND COLIN SUMMERHAYES

After studying physics and mathematics at university, Peter Dexter was manager of the marine programme at the World Meteorological Organization (WMO) for 20 years from 1984, and the WMO representative to the IOC. He returned to Australia in 2004 where he now heads the Oceanographic Services Group in the Bureau of Meteorology. Since 2005 he has been co-president of the Joint Commission for Oceanography and Marine Meteorology (JCOMM).

Colin Summerhayes was Director of the Global Ocean Observing System (GOOS) Project Office in IOC for 7 years before joining the Scientific Committee on Antarctic Research (SCAR) as Executive Director in 2004. There he helped develop observing systems for sea ice and the Southern Ocean.
He is now an emeritus Associate of the Scott Polar Research Institute.

Introduction

Although humans have been observing and recording aspects of the oceans since the beginning of recorded history, until very recently this has been done in the context primarily of increasing our understanding of the oceans and their contents, rather than for more pragmatic reasons. There were exceptions, of course, such as Benjamin Franklin's mapping of the Gulf Stream, as an aid to navigation (Figure 7.4); but by and large, oceanographers, unlike their meteorological colleagues, have been concerned more with their science than with the applications of this science and with an associated reluctance to share data with others. However, driven initially by the need to understand and predict global climate and climate change, coupled with a realization of the critical role of the oceans in the global climate system, recent decades have seen a significant increase in the requirement to systematically monitor ocean behaviour, with the observational data being widely and freely exchanged.

In response to calls from the Second World Climate Conference (Geneva, 1990) the Intergovernmental Oceanographic Commission (IOC) created the Global Ocean Observing System (GOOS) in March 1991. The creation was also a result of the desire of many nations to gather the information required to improve forecasts of climate change, the management of marine resources, mitigation of the effects of natural disasters, and the use and protection of the coastal zone and coastal ocean. The call to create and develop a GOOS was reinforced at UNCED in Rio de Janeiro, in 1992.

This paper describes the GOOS, and touches on its relation to the Global Climate System (GCOS), for which the GOOS provides the ocean component; to the Global Terrestrial Observing System (GTOS), which interfaces with the GOOS in the coastal zone; and to the more recent Global Earth Observation System of Systems (GEOSS). It summarizes the GOOS's plans, priorities and requirements, and assesses the benefits accruing nationally from international involvement in the GOOS.

What is GOOS?

GOOS is:

- a sustained, coordinated international system for gathering data about the oceans and seas of the Earth;
- a system for processing such data, with other relevant data from other domains, to enable the generation of beneficial analytical and prognostic environmental information services; and
- the research and development on which such services depend for their improvement.

Implementation of the GOOS is expected to lead to an increase either in wealth (for instance through reducing the costs of certain activities) or in well-being (by ensuring safety of life and security of property). Most countries expect to benefit from the information from the GOOS, given that a significant part of world economic activity and a wide range of services, amenities and social benefits depend upon efficient management of the sea. For many countries, marine resources and services provide 3–5% of Gross National Product. For a few it is much higher. In addition, because the ocean plays a significant role in climate change, information from the GOOS is expected to contribute to other sectors of the economy including the management of water resources, energy resources, transport, agriculture and forestry.

There are two components of the developing GOOS system, one dealing with climate and the open ocean, and the other with coastal marine observations. They are

Figure 11.1 The first long-term systematic marine data measurements were made by lighthouse keepers, often in remote places. Image by David Pugh.

part of the same system and many of the observations are common, but their goals have been separately defined.

The open ocean component of the GOOS is designed to:

- monitor, describe and understand the physical and biogeochemical processes that determine ocean circulation and its effects on the carbon cycle and climate variability;
- provide the information needed for ocean and climate prediction, including marine forecasting;
- provide observational requirements;
- ensure that the designs and implementation schedules are consistent and mutually supportive and working as planned; and
- ensure that the system benefits from research and technical advances.

Coastal GOOS has six goals for the public good. These are to:

- improve the capacity to detect and predict the effects of global climate change on coastal ecosystems;
- improve the safety and efficiency of marine operations;
- control and mitigate the effects of natural hazards more effectively;
- reduce public health risks;
- protect and restore healthy ecosystems more effectively; and
- restore and sustain living marine resources more effectively.

Plans, priorities and requirements for the open ocean

Real or potential applications linked to societal needs should drive the ‘shape’ of the requirements for the open ocean observing system for climate. Ocean data are important in their own right and via their contribution to increased forecast model skills and to the continual improvement and validation of forecast models. Among the applications are: numerical weather prediction, ocean and climate prediction, short-range prediction of ocean waves, sea-ice monitoring and prediction, and future climatologies. Operational applications, on time scales of days to weeks, include coastal and offshore engineering design, tactical ocean forecasting related to national defence and civilian protection, shelf and coastal predictions and predictability, information for off-shore industries, safety and search and rescue. Climate prediction involves understanding links between oceans and atmosphere, for example in phenomena such as the El Niño Southern Oscillation (ENSO), monsoons and the longer term variability of the Indian Ocean Dipole, North Atlantic Oscillation, Pacific Decadal Oscillation, Atlantic Tropical Dipole and Indian Equatorial Dipole, which involve global scale interactions and deep ocean circulation.

The sampling requirements for these applications developed in the late 1990s included both *in situ* and satellite observing platforms. The GCOS established a list of the Essential Climate Variables that are both currently feasible for global monitoring implementation and have a high impact on the requirements of forecasting climate change. For the oceans these include: sea surface temperature, salinity, sea level, sea ice, ocean colour for biological activity; subsurface temperatures, salinities, currents, nutrient and carbon dioxide concentrations, and phytoplankton distribution.

The systems for open ocean measurement under the aegis of the GOOS were initially designed by the Ocean Observing System Development Panel, refined in the 1998 Action Plan for GOOS/GCOS, and further refined in the GCOS Implementation Plan. Status maps of the distribution of individual components of the network, such as locations of tide gauge stations, drifting buoys, ships' tracks and so on can be accessed from the JCOMM *in situ* Observing Platform Support centre (JCOMMOPS): http://wo.jcommops.org/cgi-bin/WebObjects/JCOMMOPS. The marine meteorological observations include measurements of temperature, air pressure, wind speed and direction, precipitation and water vapour.

Much remains to be done to make the GOOS fully operational for climate forecasting. We require global coverage; more high-quality observations; ocean biogeochemical variables; ocean analysis and reanalysis combined with data assimilation to realize the value of the networks; national centres or services dedicated to implementation and maintenance of the observing system; long-term funding commitments; national operational ocean services analogous to national weather services; and commitments to sustained data streams from experimental satellites. At the same time some real progress is being made, as evidenced by the success in implementation of the planned network of Argo profiling floats in a relatively short time period (Figure 11.2).

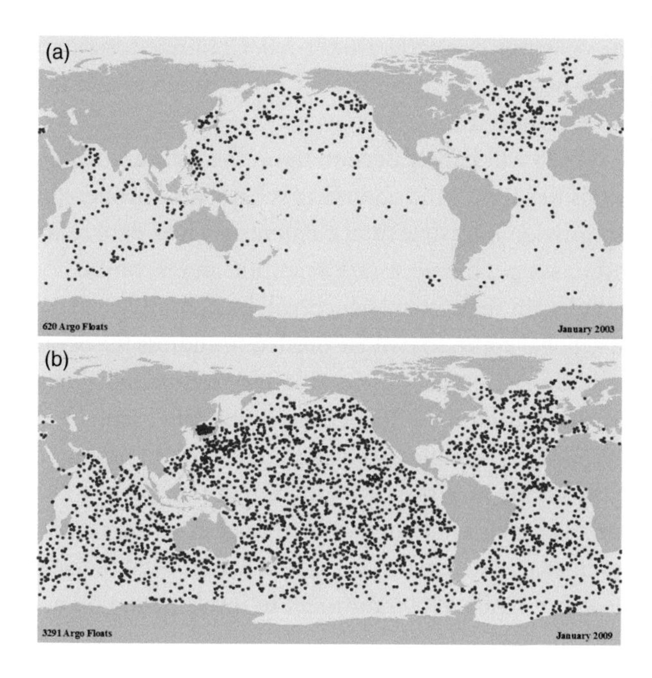

Figure 11.2 (a,b) Progress made in the Argo profiling float network since January 2003. Data from JCOMM.

Systematic sampling of the global ocean subsurface for temperature and salinity to characterize ocean climate variability and ocean behaviour will be addressed through developing the Upper-Ocean Network (currently, 3000 Argo floats; 41 Repeat XBT Lines; 29 Surface Reference Moorings; 120 Tropical Moorings; high-resolution satellite altimetry). The global reference moorings network will provide essential reference-quality, long-time records of subsurface variables to identify climate trends, and the basis for testing models. Deep-ocean time series observations are also essential for determining long-term trends. Much of this work is taking place through the Joint WMO/IOC Technical Commission for Oceanography and Marine Meteorology (JCOMM), which is the implementation arm of the GOOS.

The routine generation of ocean climate products will provide essential ocean data for global climate and weather models needed by governments. All actions are technologically feasible and can be accomplished with established coordination mechanisms and agreements. The composite approach, with carefully balanced contributing systems and networks, and broad coordination and cooperation, is essential for meeting the requirements.

The major weaknesses of the present observing system are the lack of global coverage of the ocean, the need for autonomous and remote instruments for the entire oceanic variables list, the need for expanded and more effective data and product systems, and the need for long-term continuity of national efforts along with international coordination. Present global effort depends heavily on the efforts and funds of the research community. National and regional agents of implementation with clear tasking and adequate on-going resources are needed. Continued research is required to improve and develop observing capabilities for some variables and to make systems more robust and cost-effective.

The implementation of the GOOS will largely depend upon the commitments made by the participating nations to support the observing subsystems through their national observing agencies, to provide infrastructural elements such as data centres and distribution networks, and to supply the necessary scientific and technical research and development to underpin the system. Much will be accomplished through regional alliances, but a global approach will be needed to address the ocean's role in the climate system. In addition a number of scientific and technical challenges must be met. We need satellite observations with higher resolution and accuracy and more spectral bands from geostationary satellites; we need improved observing system evaluation and design, including improvements in air–sea flux parameterizations; better ocean platforms are required; better ocean sensors and systems are wanted; and better instruments are needed. These are all technologically feasible, and are happening now, but national commitment to implementation and long-term maintenance remains the key.

Observing system requirements for the coastal ocean

The coastal ocean includes all coastal waters, including lagoons and estuaries, as well as the waters of the continental shelf and Exclusive Economic Zone. As for climate-related observations, real or potential applications linked to societal needs should drive the 'shape' of coastal applications.

The establishment and maintenance of such a user-driven, sustained, integrated and end-to-end system is the purpose of the coastal module of the GOOS. Two kinds of plans have been produced for observations from the coastal ocean. The first concerns the actual coastal ocean measurements that should be made. The second concerns the coastal ocean observing systems that should be used to make those measurements.

Integration across the land–sea boundary is essential, because changes in the state of coastal ecosystems commonly depend heavily on changes on the land, for example, increased population and changing land use can increase pressure on coastal seas. This can be through extraction of living resources, or nutrient or contaminant loading, which may lead to loss of commercial fishing value, decreased public health or costly coastal floods. These in turn require operational, usually government, responses such as fishery management, sewage treatment plants and controlled land use.

Figure 11.3 Marine mammals carrying sensors are excellent collectors of data. Image courtesy of John Gunn, Australian Antarctic Division.

Informed decisions by coastal managers, nationally or through local government agencies, demand that information is supplied at rates tuned to the timescales at which decisions have to be made. To satisfy that requirement the coastal module of GOOS must meet these criteria:

- measurements, data streams and analyses required by user groups are sustained, routine, guaranteed, continuous or repetitive as needed, and of known quality;
- measurements and data analyses (e.g. modelling) are efficiently linked via integrated data management and communications;
- observations capture a broad spectrum of variability in time, space and ecological complexity;
- observations are multidisciplinary and the resulting data streams and products support a broad diversity of applications; and
- the system provides data and information required to relate changes in environmental systems to changes in socio-economic systems.

Table 11.1 User Groups for data and information from the coastal ocean

Shipping	Marine energy	Marine mineral extraction	Insurance and re-insurance	Coastal engineering
Fishing community	Agriculture	Aquaculture	Hotel and catering	Consulting companies
Search and rescue	Port authorities and services	Weather services	Government agencies for environmental regulation	Freshwater industry
Public health authorities	National security	Wastewater management	Coastal management	Emergency response agencies
Tourism	Conservation groups	Seafood consumers	Recreational swimming and boating	News media
Educators	Scientific community	Charting and navigational services		

Present and potential users for data and information about the coastal ocean are summarized in Table 11.1. These users are interested in knowing about a wide range of phenomena (Table 11.2).

Considering the variety of user communities and their interests, six broad categories of societal issues can be identified:

- improving the capacity to detect and predict the effects of weather and climate change on coastal ecosystems;
- improving the safety and efficiency of marine operations;
- more effectively controlling and mitigating the effects of natural hazards;
- reducing public health risks;
- more effectively protecting and restoring healthy ecosystems; and
- more effectively restoring and sustaining living marine resources.

These six issues have common requirements for data and information. For example, search and rescue, health risks of exposure to pathogens while swimming, and forecasts of the trajectories of oil spills and harmful algal blooms all require nowcasts of surface current fields.

The main elements of the Coastal Ocean Observing System are:

Table 11.2 **Examples of phenomena of interest from the coastal ocean**

Sea state	Forces on structures	Coastal flooding	Currents	Sea level
Shoreline change	Seabed topography change	Chemical contamination of seafood	Human pathogens in water and shellfish	Habitat modification and loss
Eutrophication, oxygen depletion	Change in species diversity	Biological response to pollution	Harmful algal bloom events	Invasive species
Water clarity	Disease and mass mortality in marine organisms	Chemical contamination of the environment	Harvest of capture fisheries or of aquaculture	Abundance of exploitable living marine resources

- measurements (sensing), sampling and data transmission (monitoring);
- data management and dissemination; and
- data analysis and modelling (including data assimilation).

The measurement subsystem samples the ocean both remotely and *in situ* and consists of the mix of platforms, sensors, sampling devices and measurement techniques needed to measure variables on required time and space scales. The data management subsystem links observations to the data analysis and modelling subsystem and ultimately to products and services.

The coastal measurement subsystem comprises national, regional and global entities. National GOOS Programmes can bring together observations made by a range of national agencies. More broadly, nations sharing a common sea area are encouraged to work together through GOOS Regional Alliances (GRAs) to create a whole that is greater than the sum of its parts, and from which they can all benefit. Aside from GRAs, other regional bodies may have regional interests in GOOS. The International Council for the Exploration of the Sea (ICES) has an IOC/ICES GOOS Working Group to consider how ICES may contribute to the development of the GOOS in the North Atlantic and the seas around Norway. The Scientific Committee on Antarctic Research (SCAR) and SCOR are developing the design for a Southern Ocean Observing System (SOOS), which could form the basis for a Southern Ocean GRA.

A global approach is needed to recognize that many phenomena of interest occur in coastal waters worldwide and are often local expressions of larger-scale forcing. The Global Coastal Network (GCN) is designed to capture variability and change across global to local scales. The observing and data management subsystems of the GCN measure and process the common variables required by most regional systems. The GCN is essential for comparative analyses of changes occurring within regions on global scales, for global assessments of changes in the state of coastal marine systems, and for serving global products on national to regional scales. GCN will include all relevant satellite measurements.

Many of the required elements are already being monitored (often at a higher frequency than is required for the GOOS) by national agencies in industrialized countries. Not all measurements needed for coastal seas are made in the ocean. Stream gauges in rivers make an essential contribution to understanding the coastal ocean. *In situ* measurement networks are generally sparse, making the coastal ocean undersampled. Increasing coverage by such instrumented networks is a significant challenge, as is a strategy for putting multiple instruments on single measuring platforms to maximize sampling returns. There is also a significant challenge in providing continuity of measurement from *in situ* coastal systems, because local conditions (high waves, fouling, ice) may lead to demands for frequent maintenance and/or replacement of instruments.

Many coastal data are acquired these days, and probably increasingly in future, by remote sensing from satellite. High accuracy is obtained from those systems used for physical mapping of bathymetry, topography and shoreline position. Nevertheless, it is important to ensure access to the highest resolution Digital Elevation Models to determine elevation with land–sea continuity. Some sensors have problems in the coastal ocean. For example, although satellite-derived sea surface temperature is a mature and robust measurement, it requires improved resolution in space and time to address coastal needs. Although there are numerous multi-spectral satellite sensors (MODIS, MERIS, etc.) that can provide optically derived products such as pigment and dissolved or suspended matter concentrations, they are primarily focused on the global domain, and have limitations in spatial and/or temporal resolution in coastal seas. Improved instrument calibration and validation remain pressing needs. Continuity of satellite measurement is an ever present concern here and in the open ocean.

Significant obstacles exist to hinder the data integration needed for the holistic study of the coastal domain. These include communication difficulties between the disparate elements of the coastal community, data access and management issues, and the unique challenges of the coastal regions. These difficulties can be resolved through community efforts, following a national or regional plan.

Figure 11.4 Marine observations begin at the shore and extend to the deep ocean: (a) met station and lighthouse © Yegor Korzh/Shutterstock.com; (b) Woods Hole Oceanographic Institution (WHOI)-operated submersible *Alvin*, image courtesy of Woods Hole Oceanographic Institution.

Coastal GOOS will evolve in response to user demands for data and data-products and services. Evolution will depend on implementation of known technologies, and on research and pilot projects to develop new observational capabilities, especially for those goals related to public health, ecosystem health and sustainable resources. Transition from sustained observations supported by research funds to those supported by operational funds is a major step and the joint responsibility of both research and operational communities. Many of these steps are likely to be seen as desirable not merely to meet the demands of the GOOS, but also to meet the growing demands of government departments.

Global, regional and local trends in natural processes and human demands jeopardize the ability of coastal ecosystems to support commerce, living resources, recreation and habitation. Concerns have led to numerous international agreements that require sustained and routine observations of both coastal terrestrial and marine systems. Meeting the terms and conditions of these conventions and action plans requires the establishment of an integrated global system of observations for the atmosphere, oceans and terrestrial systems as part of the Global Earth Observing System of Systems (GEOSS).

The polar oceans

The Arctic and the Southern Ocean are key elements of the Earth system, since it is there that cooled water sinks to ventilate the global ocean, carrying climate signals

Figure 11.5 Research ship deployment of a drifter release. © JCOMM.

worldwide. These regions are highly sensitive to global warming. What happens there will have a considerable influence on climates elsewhere, especially at mid latitudes. Despite their importance, polar seas are poorly monitored, not least because of their harsh conditions and geographical remoteness. Many more observations of the polar oceans are needed to understand and predict climate change accurately. A key problem concerns measurement of ocean properties year-round beneath the sea ice. Modified Argo floats, moorings, gliders and instrumented marine mammals provide various means of escaping from this constraint.

The Future

Several nations have considered ways in which benefits may accrue from ocean observations. The benefits are shown to be considerable. It appears from these and many other similar studies that an ocean observing and forecasting system will generate positive dividends significantly greater than its costs; that the benefits will be significant, and that because expenditure and incomes for the various parts of the system do not occur in the same places, or agencies, or at the same times, a national and/or regional view is required to maximize the net benefits in terms of public good. The short-term products and benefits should provide an economic return that covers the investment required for the long term.

Most operational oceanography is carried out locally, to solve local problems – for instance to provide information for oil platform operators in a specific area, or to monitor and model water levels in a particular port and its approaches.
However, local conditions are always subject to regional controls, set in a global ocean-atmosphere-ice system with teleconnections between far flung areas. The benefits to any particular nation from international involvement in GOOS arise

Figure 11.6 A CTD-rosette containing 36 seawater samples is retrieved in the Southern Ocean. Image by Brett Longworth (WHOI)/IOC.

on two scales – on the broad scale and on the local scale. On the broad scale it is axiomatic that ocean processes know no boundaries. The water masses moving through a nation's Exclusive Economic Zone may come from many different parts of the particular ocean basin, including the polar regions. Given the global scale of these features, it is simply not feasible for any one nation to monitor these different water masses and the ways in which they change along their course, in such a way as to provide a sound basis for forecasting national environmental or climate changes. The task is global and demands cooperation and coordination.

Aside from providing the benefits in the form of a wide range of direct products and services (open ocean data from all sources, and integrated by national agencies), also provide the boundary conditions for those agencies running numerical models of the behaviour of their coastal waters. These waters are themselves an integral part of the coastal waters of the whole region. The sharing of data on these waters

Figure 11.7 A spray-glider returns from a mission. Spray is an underwater glider developed under ONR support by Scripps and Woods Hole scientists to provide a small long-range autonomous platform for long-term ocean measurements. © IOC/UNESCO.

is essential if the detailed behaviour of coastal seas is to be fully understood and forecast, as the basis for the provision of local products and services. Benefits come from the free exchange of data and information between neighbours, and from implementing local, regional and global ocean observing programmes in a coordinated and cooperative way.

To improve those benefits requires improving the observing system in consultation with representatives of the full range of data suppliers and end users, which together form the 'stakeholders' in the system. The stakeholders in the GOOS are much broader than organizations with purely ocean and coastal interests, because ocean data are useful **not only** in the provision of ocean products and services, **but also** integrated with data from atmospheric and terrestrial observing systems, in the provision of environmental products and services required to address the environmental concerns of many land-based sectors.

There is no doubt that the GOOS has come a long way since the initiation of the concept, more than two decades ago. For one, the concept is now part of the

Figure 11.8 Observations from space cover vast areas in a short time. Here the progress of a tropical cyclone in the Atlantic Ocean is being monitored. © Vladislav Gurfinkel/Shutterstock.com.

international environmental vocabulary, and the GOOS has an established role as the ocean component of the GEOSS. Governments, agencies and individuals generally recognize that they stand to gain many times more from cooperation and sharing through the GOOS than their individual contributions. There is also broad acceptance of the need for composite, integrated observations, and that no single approach to ocean observation can ever provide all the answers. At the same time, technological advances in observing sensors and platforms (*in situ* and remote sensing), as well as in data communications and processing, mean that we are now able to take a snapshot of the physical state of the global ocean, in real-time, on a daily basis. Combining this with enhanced scientific understanding, modelling and data assimilation techniques, and with super-computing power, we have ocean prediction models which show skill comparable to that of numerical weather prediction in the early 1980s. There is no doubt that our scientific and technological capabilities will continue to advance, and will soon extend to coupled physical and biogeochemical processes. The need for and potential of the GOOS is perhaps greater now than when we first started.

But there remain problems. Unlike in meteorology, where national requirements for data and services were (and still are) driven largely by aviation and public safety, there is yet no single, 'big-ticket' issue to drive the GOOS, apart perhaps from climate. This means that Governments see no compelling reason to invest heavily in observing the open ocean, a global commons, and a large proportion of open ocean GOOS remains funded through, and maintained by, research programmes. On the other hand, in coastal GOOS, issues of national sovereignty, security and economic interest in many cases continue to override the potential value of international cooperation and data sharing. In summary, as noted in a recent review of GOOS 'funding commitments are inadequate and mostly short-term, a problem which is particularly challenging for implementation of coastal GOOS since most of the global coastal ocean is in the Exclusive Economic Zones of developing countries'.

Figure 11.9 Mapping the earth from space is very cost-effective, and produces vast amounts of data. © Snaprender/Shutterstock.com.

Nevertheless, we do have reason to be optimistic. In many countries, drivers for the GOOS related to global change and the increasing degradation of the marine environment are gaining public as well as scientific attention. Couple this with the ever-improving skills in modelling and predicting the state of global, regional and local ocean conditions, as well as new ways of managing and displaying ocean data, and we have the means to attract and retain the attention of Governments on the need to systematically monitor the oceans on an ongoing basis – for which the GOOS provides the established and successful mechanism.

12 Oceanographic data: from paper to pixels

IOURI OLIOUNINE AND PETER PISSIERSSENS

Iouri Oliounine holds a doctorate from St Petersburg University in Oceanology and Marine Dynamics. He has worked extensively in the St Petersburg Arctic and Antarctic Research Institute, and in the Russian Hydromet. He has also been Deputy Executive Secretary of the IOC, and Executive Director of the International Ocean Institute, Malta.

Peter Pissierssens obtained a degree in Marine Sciences from the Free University of Brussels. He started his international career in 1985 in Kenya. In 1992 he moved to IOC, Paris where he has responsibilities related to marine information management, data management (IODE), and the tsunami warning and mitigation programme (ITSU). In 2008 he was appointed Head of the IOC Project Office for IODE in Ostend, Belgium.

Oceans need science and science needs data. Without proper measurements to guide and control theories, everything and nothing are possible. As Andre Gide says '*On carelessly made or insufficient observations how many fine theories are built up which do not bear examination*'. Governments need tested theories and valid forecasts. The oceans are vast and observations are expensive, yet the need to monitor and understand changes to our planet that is two-thirds covered by oceans is increasingly a human imperative.

Over the past 50 years scientists and governments have more and more appreciated the value of archiving and having access to the expensively collected marine data and information. By exchanging data among data collectors and users, much more data can be accessed, but progress towards open data exchange has been slow and spasmodic, often driven by external urgencies such

Figure 12.1 The analogue computer based on a system of pulleys and wheels, used for predicting tides until the 1960s. © Proudman Oceanographic Laboratory.

as the 2004 tsunami and earlier disasters, and by the sustained commitment of a few dedicated individuals. From the early days of data, recorded on paper and exchanged by slow postal systems, to the present, digital power of personal computers and instantaneous access through the Internet, the leader in marine data exchange has been the International Oceanographic Data and Information Exchange (IODE) of the IOC. This chapter records the work of the IODE in the wider marine context.

Marine data underpin many of the activities we undertake: in scientific research, modelling, monitoring and assessment. These data are precious; they are fundamental to the understanding of the processes that control our natural environment. The data help provide answers to both local questions, such as the likelihood of coastal flooding, and global issues, such as the prediction of the impact of global warming. The better we can predict these events, the better we can protect ourselves in the future. This not only affects everyone today, but also, potentially, the quality of the lives of future generations. The activities described in this chapter form a necessary part of the process to meet these aspirations.

It all started in October, 1961 at the First meeting of the Intergovernmental Oceanographic Commission (IOC) when Member States launched the programme on oceanographic data management, recommended the establishment of National Oceanographic Data Centres (NODCs), decided to establish an intergovernmental working group to assist Member States and to coordinate efforts. This agreement was a major triumph for ocean science at a time when the East–West cold war was at its peak.

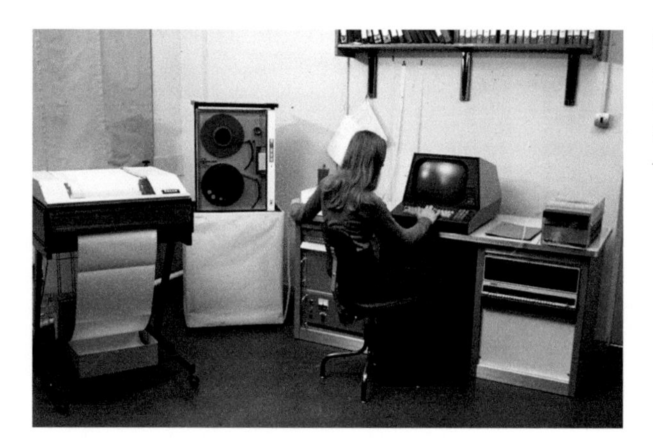

Figure 12.2 This 1980s computer used magnetic tape, and produced vast quantities of paper printout. © Proudman Oceanographic Laboratory.

The long-term and successful practice of exchanging oceanographic data had already begun on a regional basis through the International Council for the Exploration of the Sea (ICES), established in 1902. This regional initiative indicated that the creation of a global system for ocean data exchange would have great potential. In a wider context, the International Geophysical Year (IGY) (1957–1958) had recently finished and a major study on the Indian Ocean (the International Indian Ocean Expedition – IIOE (1959–1965)), jointly planned by IOC and the Scientific Committee on Oceanic Research (SCOR) was underway. The improvement of economic conditions worldwide made it possible to plan scientific oceanographic expeditions with the participation of many states. The IGY marked one of the first attempts to implement a cooperative approach in ocean research. In addition, as a follow up to the IGY, the system of World Data Centres, as a final repository of data, was established by the International Council for Science (ICSU) and included centres for oceanographic and marine geology and geophysical data. However, in the area of data management the IGY and IIOE had raised as many questions as they answered such as: What should be done in order to have accurate, reliable and readily available data? How should data be made easily available to users? What policy should be developed to encourage data holders to share the data?; and many related issues. . . .

Development of the IODE system

From the very beginning the IODE was considered as an overarching programme interrelated with all other activities of the IOC. It was evident that reliable decisions or successful planning can only be achieved when there are data and when these data are accessible and reliable. This requirement posed significant demands on the quality

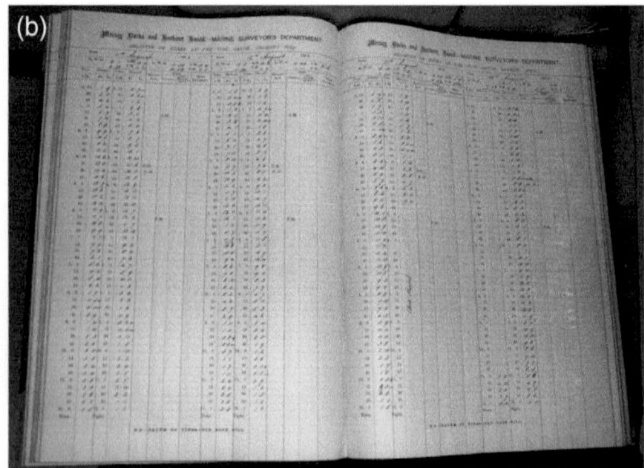

Figure 12.3 (a,b) To recover the data from these old tide gauge records is both expensive and tedious to recover. However, the old data are invaluable for studying long-term sea level changes. Each sheet records 1 or maybe 2 weeks of sea levels. Images by David Pugh.

and quantity of marine and coastal data collected as well as on the availability of data for their use.

The main objectives of the IODE Programme at the time of its establishment were to: (1) facilitate and promote the exchange of oceanographic data and information; (2) develop standards, formats and methods for the global exchange of oceanographic data and information; and (3) assist Member States in acquiring the necessary capacity to manage oceanographic data and information and become partners in the IODE network.

Established as an IOC working group, IODE started to implement its terms of reference by organizing the first session in 1962 in Washington DC, USA, where national and world oceanographic data centres were already established. The chair of the session was Dr W. C. Jacobs from the United States who was one of the strong supporters of the establishment of an international data management system.

The first years were those of trial and error when the IODE was in search of its unique place among existing data management systems of other organizations such as the World Meteorological Organization (WMO) and ICSU. It took time to understand that data management is a collective responsibility shared by all, that there is a need to develop a complex set of regulations to govern the system, and that there was a need for a much stronger and efficient intergovernmental structure than just a working group. Sometimes the IODE was too slow in responding to new requirements formulated by scientists and other users, or in catching up with the rapid development of technology for data collection and exchange, analysis and product development.

Figure 12.4 These printed publications from ICES in the 1970s were the standard way of storing and transmitting data. Image by David Pugh.

At the Eighth Session of the IOC Assembly in November 1973, the Working Group was renamed as a Working Committee on International Oceanographic Data Exchange and the terms of reference were modified to better meet new needs for ocean and coastal data. New data types were incorporated to include not only physical and chemical data but also geological–geophysical, biological, air–sea interaction, marine pollution data and data from continuously recording and remote sensors.

The IODE also established groups of experts to undertake detailed scientific and technical studies and to advise the Committee. The membership of the groups includes not only experts drawn from the IOC Member States, but also representatives of other IOC partner organizations with experience in data or information management, such as the ICSU, ICES, WMO, Food and Agriculture Organization (FAO) and others.

In 1987 the IOC changed the name to the 'Committee on International Oceanographic Data and Information Exchange' to raise the profile of information for achieving easier access to the data and considering mechanisms for the exchange of bibliographic and scientific information. The IODE acronym was retained.

A policy for ocean data exchange

Governments, like individual people, are careful about giving something away without a fair return. From the outset is was clear that data could be made available and useful only if there were internationally agreed practices and associated institutional arrangements for the exchange of oceanographic data based on the principle of timely full and open access to quality ocean data. The principle was articulated by the IODE in cooperation with the ICSU Panel on World Data Centres in 1991 and presented in the 5th edition of the Manual on International

Oceanographic Data Exchange published jointly by the IOC and ICSU. The policy helped to facilitate efficient and appropriate utilization of data, was flexible enough to permit effective new partnerships and was responsive to new priorities. The aim was to maximize the use and value of the data.

This early vision was sometimes threatened by more hard-line approaches. Throughout the 1990s, an attempt was made to change the existing IOC–ICSU policy towards data commercialization, which could have far-reaching implications and consequences for marine science programmes. With the active participation of IODE experts, the IOC Member States, subsidiary, technical and regional bodies, reviewed and assessed the implications of the proposed modifications. The result was an agreement to keep in force the main IODE–ICSU principles, and to confirm the commitment to open ocean data exchange.

More recently, an important milestone was achieved when agreement was reached by Member States on an IOC Oceanographic Data Exchange Policy. The preamble to the Policy, adopted by Member States in 2003, affirms that *'the timely, free and unrestricted international exchange of oceanographic data is essential for the efficient acquisition, integration, and use of oceanographic observations gathered by countries of the world for a wide variety of purposes including the prediction of weather and climate, the operational forecasting of the marine environment, the preservation of life, the mitigation of human induced changes in the marine and coastal environment, as well as for the advancement of scientific understanding that makes this possible'.* It leaves open the question of charging for products derived from the basic data, while encouraging free exchange wherever possible.

Training and education in data management

Partnership in exchanging expertise, as well as in exchanging information and data, remains one of the core objectives of the IODE programme. One reason for this strong focus on capacity building has been the lack of formal education and training available in the field of oceanographic data management. In general, expertise was passed within the NODC from generation to generation. The recognition of data management as a specialized profession was slow in developing, and the skills were not generally available. Through the IODE network the IOC succeeded in sharing this valuable experience beyond the institutional level. Many training courses and workshops were organized, missions to the regions and Member States were undertaken, and internships provided to assist in the establishment and operation of NODCs and Marine Information Centres.

In the 1980s, in an effort to standardize training courses, to provide resources for capacity building and utilize personal computers, which were becoming increasingly prevalent, the OceanPC software package was developed and widely used. This

Figure 12.5 (a,b) Data centres around the world still have vast quantities of ocean information on their shelves. Images by David Pugh.

provided tools for ocean data entry and quality control of temperature, salinity and nutrient data. OceanPC was developed by Harry Dooley (ICES) and Murray Brown (USA).

In the late 1990s this concept was expanded and the IODE Resource Kit was developed to assist with training. This CD-ROM based product comprised a broad range of marine data and information management material, including software, quality control and analysis strategies, training manuals and relevant IOC documents. It was designed to be a comprehensive self-training and resource tool, for newly established oceanographic data centres, and to assist managers and staff members to acquire the skills to set up and run new IODE centres. This in turn evolved into OceanTeacher – a modular, expandable, web-based educational tool, based upon an encyclopaedic resource module and training curricula. Further development is effected under the name of OceanTeacher Academy.

Marine information management

The Marine Information Management component of the IODE was created to promote the availability of marine information and to develop tools supporting this objective. A variety of information has been gathered and exchanged: inventories of available data were developed; lists of specialists working in the area of aquatic and marine sciences were compiled at the national and regional levels and put to use; and exchanges of scientific and technical publications were facilitated.

Through the work of the IODE Group of Experts on Marine Information Management (GE-MIM), many useful tools were developed, including the Global

Directory of Marine and Freshwater professionals (OceanExpert), the catalogue of ocean-related websites (OceanPortal) and the e-repository of full-text ocean publications (OceanDocs). GE-MIM also established a strategic alliance with the International Association of Aquatic and Marine Science Libraries and Information Centres (IAMSLIC). In addition, the IOC/IODE is a partner in the Aquatic Sciences and Fisheries Abstracts (ASFA).

Working in partnerships

The success of the IODE was to a large degree related to the development of close partnerships with many international organizations such as the International Council for Science (ICSU), the International Council for the Exploration of the Sea (ICES) and the World Meteorological Organization (WMO) in addition to working among governments, industry and the marine community. These partnerships, both short- and long-term, have been developed to meet the needs of multidisciplinary science programmes, to avoid duplication of effort, to design common methods of data and information management and provide an opportunity for sharing the work with resulting savings in resources.

During the second half of the 1980s, scientists planning major research programmes began including data management as an important component from the very beginning and invited data professionals to assist in the development of data management plans. This helped to establish a proper dialogue and closer working relations between the researchers and the data managers. Participation of the IODE in programmes of global scale such as the Tropical Ocean–Global Atmosphere (TOGA) (1985–1994) and the World Ocean Circulation Experiment (WOCE) (1990–2002) of the World Climate Research Programme clearly demonstrated the value of such an approach.

The cooperation between scientists and data managers during TOGA and WOCE represented a successful effort to make available the highest quality data in a timely manner, helping to generate readily accessible large data sets combining different data types as an input to global and ocean basin models, climate diagnostic studies and climate change assessment.

The cooperation between the ICSU World Data Centre (WDC) Panel and the IODE is a special case where two organizations, intergovernmental (IOC) and non-governmental (ICSU), have worked closely together on the publication of a joint manual on data management and exchange, including joining forces in preserving the policy of free exchange and open access to oceanographic data and information. This was acknowledged by Professor Ferris Webster (USA), the Chair of the ICSU/WDC Panel '... *the international management of ocean data under the auspices of IODE is exemplary*'. Due to the existence of the IODE network, the

Figure 12.6 (a) Punched cards were still commonly used for data entry and programming until the mid-1970s.The standard IBM card had 80 columns, each representing a digit; (b) invented by IBM, floppy disks in 8-inch (200 mm), 5¼-inch (133.35 mm) and 3½-inch (90 mm) formats enjoyed many years as a popular and ubiquitous form of data storage and exchange, from the mid-1970s to the late 1990s. Images by David Pugh.

common nature of the oceans and many decades of work by the ocean science community, marine data management occupies a leading position, ahead of many environmental activities.

More recently IODE has cooperated with the Data Management Programme Area of the new joint IOC/WMO Technical Commission for Oceanography and Marine Meteorology (JCOMM). A joint expert group on data management practices has been established together with a joint JCOMM/IODE 'Ocean Data Standards Pilot Project'. This brings even closer together data management experience from both the IOC and the WMO.

Highlights from the decades

As discussed earlier, progress has sometimes been spasmodic, but over the past 50 years of development in data management and exchange, each decade has had its own particular theme and achievements.

During **the first 20 years**, the number of national oceanographic data centres increased from seven in the first half of the 1960s to 24 by the end of the 1970s. By the year 2000 there were 47 centres. Today the IODE family comprises over 80 centres.

The continuous growth in the number of data centres was achieved thanks to a better appreciation by senior scientists and government administrators of the fact that data and information play an important role at all levels of ocean research, use and governance, but only if they are accurate, reliable and readily available. Other factors were the publication of guidelines for the establishment of a national oceanographic data centre, organization of awareness missions to the regions, for example, to the Mediterranean countries and South America. In today's language, the IODE National Oceanographic Data Centres are a good early example of the ethos 'Capture once, use many times', a pragmatic philosophy that gets the maximum value from each measurement of the ocean.

Through the **1970s**, an important part of the remit of IODE was to gather and exchange a variety of information including inventories of available data. One such tool was the Marine Environmental Data Inventory (MEDI) – a directory system for data sets, data catalogues and data inventories, with editions published in 1978, 1985 and 1993. It was set up to ensure the widest possible coverage of data holdings and included a review of existing national and international data directory systems as well as implications of interoperability with similar systems within other international organizations. MEDI was a predecessor of the European Directory of Marine Environmental Data (EDMED) and of the Global Change Master Directory (GCMD) of the USA and a forerunner of the discovery metadata systems that are in use today.

Another tool was the Report of Observations/Samples Collected by Oceanographic Programmes (ROSCOP), conceived in the late 1960s in order to provide an inventory for tracking oceanographic data collected on research vessels. The ROSCOP form was extensively revised in 1990, and was renamed the Cruise Summary Report (CSR). Most marine disciplines are represented in the CSR, including physical, chemical and biological oceanography, marine geology and geophysics, fisheries, marine contaminants and marine meteorology. Traditionally, the Chief Scientist submits a CSR to their National Oceanographic Data Centre within several weeks of the completion of a cruise.

More recently these cruise report submissions have been updated. ICES led the effort to digitize the ROSCOP information and pioneered the development of a database for this information, and, in collaboration with IOC/IODE, developed and maintained a PC-based entry tool and search facility. The emphasis for this was on ICES member countries, but it extended to other countries who wished to submit their information. The activity gained new momentum in Europe during a number of EU-funded data management projects. The CSR database now comprises details of over 35 000 oceanographic research cruises primarily from Europe and North America, but including some other regions where the information is available.

Figure 12.7 (a) The IODE Project Office in Ostend, Belgium; (b) students on a marine data management course at Ostend. Images by David Pugh and Peter Pissierssen.

In the **1980s**, a IODE Group of Experts was given the task of developing a format for data exchange that would also carry supporting information in such a way that their meaning and relationship to the observational data would be obvious. At the same time the format should be flexible enough to manage virtually any ocean measurement and any data structure (depth profiles, moored time series, underway data, remotely sensed data, etc.). Data in such a format would be easily transportable between different mainframe computers and workstations without any additional work.

As a result, during the 1980s, a general formatting system for geo-referenced data (GF3), supported by a comprehensive software interface package, was developed with a big contribution from UK experts led by Merion Jones, and experts from Canada, Russia and ICES.

The GF3 formatting system was designed as a general purpose format for exchanging oceanographic data on the media of the day, half-inch 9-track magnetic tapes. The structure of the format gave the freedom to encode virtually any type of geo-referenced data collected in the form of a data series. It was adopted in 1981 as the international exchange format for oceanographic data. The documentation for the GF3 included a set of five volumes describing the format, its uses and its supporting software and was available in English, French, Spanish and Russian.

Although there have been many changes since the 1980s, one key aspect of GF3, the code tables, which standardized many of the variables including parameter names, platform types and country names, live on today and have formed the basis of modern vocabularies and ontologies.

Data archaeology and rescue was a theme of the **1990s**. Increasing concern by governments and scientists about climate change, and the recognition of the role of the world's oceans in this process, emphasized the need for historical oceanographic data. Much historical data was still unavailable and held on obsolete or decaying media in many institutes. In 1992 the IODE launched a new project under the title 'Global Oceanographic Data Archaeology and Rescue Project (GODAR)' to improve this situation. The project initiator and leader was Sydney Levitus, Director of the World Data Centre, Oceanography, in the USA. Regional workshops, training courses, missions and meetings were held culminating in 1998 with the release of the first World Ocean Database including more than one million water bottle stations, almost 100 000 high-resolution conductivity-temperature-depth casts and almost two million bathythermograph casts. Products based on this database included the series of World Ocean Atlases with objectively analyzed fields of several oceanographic variables on CD-ROM and printed volumes. These have also been made available without restriction via the Internet. They are used by the Intergovernmental Panel on Climate Change (IPCC) and to study national and regional environmental oceanographic problems.

In some regions, for example, the Mediterranean and Black Sea, regional sub-projects were implemented. This included the MEDAR/MEDATLAS project. The objective of this project has been to rescue, safeguard and make available a comprehensive data set of oceanographic parameters collected in the Mediterranean and Black Sea. The MEDAR/MEDATLAS produced a set of CD-ROMs comprising a research cruise inventory and software for quality control in addition to a database of observed data (nearly 300 000 temperature profiles, over 100 000 salinity profiles, almost 50 000 dissolved oxygen profiles and other variables such as pH, nitrite, phosphate, chlorophyll) and a Mediterranean and Black Sea climatology. The project leaders were Catherine Maillard from France and Ethstathios Balopoulos from Greece. This GODAR regional project helped to safeguard a large number of marine datasets held by the bordering countries of the Mediterranean and the Black Sea as well as improving the data flow and communication between the data holders in the region.

Development of regional networks worldwide was a focus of **2000s**, based on pioneering experiences in Eastern Africa. As early as 1987 the IODE began considering the development of regional networks. A feasibility study in the Eastern African region was undertaken to establish a mechanism for exchange of bibliographic and scientific information. With a regional hub based in Mombasa, Kenya, a successful network and service was developed providing thousands of bibliographic references and documents to ocean researchers in eight countries in Eastern Africa.

Figure 12.8 Data stored on obsolete media may be impossible to read in the future. Image by David Pugh.

The project created not only institutional linkages but also human 'informal' networks between hundreds of researchers. This led countries in West Africa to develop a similar network for their region. The countries in the Eastern Africa region took steps to extend the data network. Implementation started in 1997 with the Ocean Data and Information Network for Eastern Africa (ODINEA), which expanded into a Pan-African network: the Ocean Data and Information Network for Africa (ODINAFRICA-II).

The success of ODINAFRICA resulted in this capacity building model being widely used. It is based on four guiding principles: (1) linking training, equipment and operational support; (2) regional context: focus on national requirements but also identify similar needs across a region; (3) product and service oriented: develop data centres not as isolated facilities, but as centres that provide the services and products that are needed by national and regional users; and (4) multi-stakeholder approach: ensure that the project is driven by stakeholders as representatives of users and involve these stakeholders in the governance of the project.

Subsequently, similar networks in other regions have been developed: Latin America and the Caribbean (ODINCARSA, started 2004); Central Indian Ocean (ODINCINDIO, 2005); European Countries in Economic Transition (ODINECET, 2007); Black Sea region (ODINBlackSea, 2007); western Pacific region (ODINWESTPAC, 2007); and Regional Network of Pacific Marine Libraries (ODIN-PIMRIS, 2008).

A project office for data marine exchange

The 25 April 2005 was a special day in the history of the IODE. An IOC Project Office for IODE was officially inaugurated in Ostend, Belgium with substantial support from the Government of Flanders and the City of Ostend. The objectives of the Project Office include: further development and maintenance of IODE and partner data and information projects; providing services and products; assisting in strengthening Member States' capacity to manage ocean data and information; and providing courses for students on data and information management.

Today the Project Office is widely known for organizing training programmes and tools and providing an environment where ocean data and information experts and students can meet. In addition, IODE ocean information systems and related public awareness tools are developed, hosted and maintained and a laboratory environment is available for the development and testing of data and information management technology. The Project Office promotes collaboration between all who are active in ocean data and information management including scientists, data managers, other IOC and partner organizations' programmes and projects, and other users.

Achievements and future challenges

The open exchange of marine data is one of the fundamental needs of both global oceanography and of more local marine management. In the past half century both the technologies of data management and the agreements for intergovernmental cooperation have changed dramatically. The key to this evolution has been through the International Oceanographic Data and Information Exchange programme of the IOC.

Since IODE was established, its data centres have acquired, quality controlled and archived millions of ocean observations, supported several international science programmes, and assembled and published many project datasets for national and regional projects. The IODE remains committed to the long-term archiving and dissemination of oceanographic data, metadata and information to safeguard present and future holdings against loss or degradation. The IODE has evolved and grown over the 50 years of the IOC in terms of the volume, variety and complexity of the data handled. Increasingly the NODCS are dealing not only with delayed-mode (for example cruise) data, but also with operational (near-real-time and real-time) data. In addition biological/biodiversity data are becoming increasingly important.

It would have been impossible to foresee all these developments 50 years ago, just as it is impossible now to anticipate progress over the next half-century. Nevertheless, some activities already underway, and their value can be anticipated.

In 2009, the Member States of the IOC agreed to assume responsibility for the Ocean Biodiversity Data System (OBIS), which had been developed as the ocean component of the Census of Marine Life (CoML, see page 147). In so doing they recognized the importance of the pioneering work on biological information and

Figure 12.9 (a,b) Colourful books but it is almost impossible to access the information they contain. Images by David Pugh.

related research and the need to ensure that the effort would continue as a legacy of the CoML. The addition of OBIS to the IODE network will also complement the available physical and chemical data bases and will facilitate the development of multi-disciplinary studies that are required to understand more fully the complexities of ocean processes and their changes.

The IODE's Ocean Data Portal (ODP) will be further developed to provide interconnection between all centres and with other networks. The ODP will provide the users with seamless access to collections and inventories of marine data from data centres and will allow for the discovery, evaluation and access to data via Web Services. To realize this vision there is a requirement to achieve inter-operability between the networks and thus reach agreements on standards. The resulting powerful data system will enhance the efficiency of interdisciplinary research, which is essential to study the relationship between climate change and environment, food and water availability, population, resources, etc., and will be of benefit to policy makers.

Ocean data managers must take a more dynamic approach in promoting the multiple uses of marine data. IODE and its data centres are ready to provide high-quality services to users, and in a timely manner. The concept of 'data publishing' needs to be more generally applied, so data sets become citable entities with the published journal article.

In the coming years capacity built through the ODIN networks must be fully utilized with a focus on national requirements and regional needs, and on providing services and products required by users. This process will involve scientists, and governments who participate in and undertake the governance of ocean management. Allied to this, the OceanTeacher Academy will adjust its direction: rather than focusing training on one region and fitting the training within the regional project, each year it will survey training requirements globally and organize courses based upon the priorities found, giving more attention to providing ocean data and information products and services required by users.

Over the past 50 years governments have been executing a variety of activities efficiently and effectively, providing global mechanisms to promote open and full access to ocean data and information. This has benefited national governments, as well as enhanced regional and global marine organizations: agencies are now undertaking activities in ocean and coastal data and information exchange which were not possible before. This will continue and expand in the next 50 years.

Our guarantee for the future is based not only on improvements in technology, but also on the mature recognition by governments that free and full exchange of marine data is a benefit to all nations. The emergence of the marine data management community, based on an environment of reliability, respect and friendship, is sometimes defined as the IODE family. This is a family of experts from different countries and organizations and members of the IOC Secretariat working together with dedication to help solve the problems in ocean and coastal data and information management.

The medieval Persian poet Sa'di of Shiraz may not have been thinking about the future of the oceans when he wrote 'whoever acquires data and does not use it resembles a person who ploughs his land and leaves it unsown'. But we have sown, and there will be a good harvest. As Sa'di also said 'A little and a little, collected together, becomes a great deal; the heap in the barn consists of single grains, and drop and drop make the inundation'.

Part V Applications: Preface

The public expects that ocean science, as is the case for all sciences, can demonstrate its usefulness in order to justify its existence, and in particular its funding. For many governments, the benefits of spending scarce resources on particular projects are expected to accrue reasonably rapidly, ideally within the period for which the government expects to remain in power. This short-term horizon is generally inappropriate, as the benefits derived from science and research usually follow many years after the initiating funding. In fact, in many instances, science programmes need to be continuous, or at least funded over long periods, for their eventual benefits to be realized. An ongoing example is research into climate change, where analyses must be based on observations stretching over decades, and where the results are needed to project many years into the future. The insidious but relentless rise in sea level due to the greenhouse effect and melting glaciers is one example where the expenditure on future sea defences will be enormous and advance knowledge of expected flood levels will be of great economic value.

For ocean issues, one must expect, eventually, a suite of benefits as a result of scientific research, covering the spectrum of local to global effects. National priorities will have first demands on a countries' scientific capacity, but due to the dynamic nature of the ocean environment, regional and global interests are almost certainly going to be involved at some stage. Again, looking at the climate change issue, the cumulative effects of anthropogenic activities are known to be serious enough to warrant global cooperation from the outset.

An immediate and compelling benefit derived from marine science is the contribution to the warning ability for extreme events, such as tsunami, storm surges and wave activity. The reaction of the international community to the Indian Ocean tsunami of December 2004 allowed a warning system to be set up, based on sound science, and a communication infrastructure put in place to mitigate against future occurrences in the area. The disaster also galvanized the governments in other parts of the globe to assess their abilities to cope with a tsunami, and to act accordingly. Unfortunately, for such events that are so sporadic and whose repetition is unlikely over long periods of time, it will be difficult to maintain the funding and interest needed to keep such systems active. Firm intergovernmental agreements are called for.

Of all the areas of the oceans important to society, the coast is the zone where the interaction with human activities is highest, and where the marine environment is at its most vulnerable. Much has been written over the past quarter century on coastal zone management; far less has been accomplished. Despite having an intergovernmental agreement on land-based sources of marine pollution, our wastes still pollute the coastal seas and our developments and land planning activities still allow the destruction of valuable coastal habitats. The coastal zone remains the area where the most immediate benefits from science are possible, but where much ignorance still exists and much capacity still has to be developed. It remains a ubiquitous worldwide problem, with global and local implications.

There are many examples of scientists working together, both formally and informally, in ocean areas around the world. We have invited two representative chapters on regional cooperation. They have significant differences: the African analysis is continental in scope, outlining issues for a vast area with its surrounding oceans and seas; the Caribbean analysis is for a regional sea containing many islands and having many surrounding coastal states. What both regions have in common is the need for governments to work together to solve their respective issues; both areas are also in need of more skill development programmes to assist them in their endeavours.

13 Life on the edge: managing our coastal zones

LAURENCE MEE

Laurence Mee is Director of the Scottish Association for Marine Science. Laurence studies coupled marine social-ecological systems. He was the UK's first Professor of Marine and Coastal Policy. Earlier, he was responsible for the $110 M GEF Black Sea Environmental Programme, led the UNEP and IAEA funded Marine Environmental Studies Laboratory in Monaco, participated in drafting the Oceans chapter of Agenda 21 and worked for 10 years in marine research at UNAM, Mexico.

Introduction

Vast lengths of our coastlines are now characterized by the heavy footprint of human development. Data projections suggest that there are currently some 3.1 billion people – 45% of the world's population – living in the 10% of land loosely defined as the 'coastal zone'; up by 765 million since 1990. The coastal human environment now includes mega-cities dotting all continents, sea defences, ports, aquaculture developments, massive plantations of palm oil trees, tourism complexes, wind farms, industrial sites, mines, military areas, roads, airports and many other features that we are gradually regarding as 'normal'. But each development has major impacts on increasingly fragile marginal natural habitats, compromising their ability to contribute to the ecosystem services upon which we ultimately depend.

The coastal zone has been a favoured place for human settlement for millennia. People have been driven to the coast by inland conflicts and resource shortages or attracted to it by the promise of a better lifestyle. The margin between coastal and marine ecosystems was generally rich in natural resources. With the development of safer maritime transport, the sea also offered enhanced opportunities for trading and a regular supply of fish. But coastal zones are also prone to natural disasters; the 2004 tsunami in Indonesia or the 2005 hurricane Katrina that destroyed major parts of New Orleans, are tragic milestones in a long history of similar catastrophic events,

Figure 13.1 Offshore wind farms are a new source of energy for the twenty-first century; and another factor in coastal planning. Image by David Pugh.

each involving more people because the coastal population continues to rise. It is also well known that coastal communities that did not remove natural fringing mangrove forests fared much better in the 2004 tsunami; a sharp reminder that the optimum value of ecosystem services may not be released through their immediate exploitation.

Such obvious levels of human pressure, coupled with increasing awareness of the pace of environmental degradation, have led to societal demands for management action. In this chapter, we will look at what has been done, appraise whether or not it has worked and consider what additional actions may be necessary.

The call for coastal zone management

Our coastlines are hugely overcrowded and the population is increasingly urban. On the Eastern Pacific seaboard, for example, 60% of China's 1.3 billion people live in coastal provinces, and Japan and Vietnam's population is almost entirely coastal. By 2015, there are expected to be some 30 mega-cities worldwide (a mega-city is defined as having a population greater than 8 million people) and 20 of these will be coastal.

Figure 13.2 Mangrove forests, such as this example from Roatan, Honduras, provide protection and an essential coastal habitat for the indigenous fish. However, many are threatened by coastal developments. © Nickolay Stanev/Shutterstock.com.

It is worth reflecting on what happens in poorly managed and regulated coastal areas: with so many people, finite resources and incompatible needs, those resource uses with low immediate value begin to be traded off against those of high value. In the absence of strong regulation, urban sprawl quickly develops because land for housing and industry commands much higher prices than for agriculture and this in turn is valued more than space set aside for conservation. Urban areas, and particularly mega-cities, demand water from surrounding areas, often at prices that easily out-compete those that can be afforded by farmers. Effluent from these areas is often disposed of in the sea or estuaries, compromising other uses, including food production or tourism. Urban services, including waste treatment, demand energy and this too eventually impacts the system by contributing to climate change and sea level rise. Demand for food puts increasing pressure on coastal fisheries, often pushing up prices and making overexploitation more attractive. It is not only the natural system that suffers from trade-offs in the coastal zone though. In many places, marginalized social groups can only afford to live on land with the lowest value and the most socially vulnerable people become the most vulnerable to natural disasters or to the health issues associated with unsanitary conditions.

Unfortunately, many aspects of the scenario portrayed above are commonplace in today's world. Efforts to manage this situation are not new; integrated planning featuring social improvement and environmental protection was applied by the British colonial administration in several Caribbean islands in the mid nineteenth century and had been part of the social norms of some traditional societies much earlier. But it was not until the early 1970s that modern integrated planning began to develop. By that time, most coastal regions of the world were afflicted by uncoordinated single sector regulations, complex planning systems and power politics between major stakeholder groups that protected their (mostly legitimate) interests. This often resulted in political gridlock leading to conflict or short-term unimaginative decisions. It is still commonplace to find that over 10 different authorities and interest groups need to be brought together in order to devise a long-term strategy for a coastal district and each sector often has entrenched positions and its own complex rules and bureaucracy. These groups might include local authorities; water boards; chambers of commerce; local branches of ministries of agriculture, environment, tourism, energy and transport; nature protection agencies; non-governmental organizations; harbour authorities; fishing cooperatives; housing associations; and many others.

The 1972 Coastal Zone Management Act (CZMA) in the USA was one of the first serious attempts to introduce a more rational and multisectoral planning legislation. This and similar early efforts in other countries led to the development of Integrated Coastal Zone Management (ICZM), a term initially coined at a conference in Charleston, USA in 1989 and generally referred to without the 'Z' (i.e. ICM) in the USA. It was defined as a: '*dynamic process in which a coordinated strategy is developed and implemented for the allocation of environmental, socio-cultural and institutional resources to achieve the conservation and sustainable multiple use of the coastal zone*'. The concept of ICZM was enshrined in 'Agenda 21', one of the main products of the UN Conference on Environment and Development held in Rio de Janeiro in 1992, the biggest gathering of world leaders in the twentieth century.

There are two important features of this definition. The first is that it is about *allocation* of resources and the second is that it sets out to achieve a *balance* between conservation and use. These are tough challenges because they assume that somebody will have the authority to allocate resources and there is a genuine commitment to the 'long view' of conserving the environment for the benefit of future generations. This requires a step change (more of a 'quantum leap' in many cases) in the way coastal environments are managed.

Figure 13.3 By 2015, there are expected to be some 30 mega-cities world-wide (a mega-city is defined as having a population greater than 8 million people) and 20 of these will be coastal. Photo of Singapore, © Oksana Perkins/Shutterstock.com.

A recent study by one of the pioneers of ICZM, Professor Biliana Cicin-Sain, summarized the key challenges for achieving ICZM as *integration* in the following areas:

- *intersectoral* (bringing together agencies and groups from different sectors);
- *intergovernmental* (national, provincial, local authority over the coastal zone and ocean).
- *spatial* (joining land and marine-based management)
- *science management* (making the best use of relevant science from all disciplines);
- *international* (particularly where there are transboundary issues involved).

There is a general understanding that this wide scope of integration can only be achieved through a combination of 'bottom-up' and 'top-down' actions. 'Bottom up' refers to the need to achieve a consensus at the local community level whereas 'top down' refers to a more directed approach from national governments. Success is therefore heavily dependent on a mix of consensus and authority.

In the 40 years or so since the concept of ICZM was first devised, there have been major efforts to implement it, to develop knowledge and tools to meet the challenges of integration, and to finance pilot studies. At least 98 coastal countries currently have some kind of programme in ICZM though the nature of these efforts ranges from small-scale pilot studies to major policy development along the entire coastline. There are major ICZM research projects and entire organizations dedicated to assist with its implementation.

Is ICZM working?

The fact that so many ICZM policies and projects have been developed across the world is a testament to the importance of this concept. There are many good examples of how ICZM practices have led to better integration between a wide range of parties to resolve conflicts, improve overall welfare and bring environmental benefits. In South-east Asia, for example, a major programme called Partnerships in Environmental Management for the Seas of East Asia (PEMSEA), partly funded by the Global Environment Facility (the GEF, a major funding mechanism that was an outcome of the 1992 Rio Conference described earlier), has led to successful full-scale local initiatives in the Philippines, Vietnam and China. Similar GEF supported initiatives have been developed as far afield as Patagonia and Russia.

Despite the major efforts to implement ICZM, there is clear evidence of continued degradation of coastal systems throughout the world. Natural habitats are shrinking, many coastal fisheries are failing – or are exploited beyond their safe biological limits – and there are widespread reports of deterioration of the human environment in many coastal areas. In the USA, for example, a recent report concluded that one-quarter of coastal lands will be urbanized by 2025, up from 14% in 1997. The report commented that this was not entirely a result of population rise: '*Runaway land consumption, dysfunctional suburban development patterns, and exponential growth in automobile use are the real engines of pollution and habitat degradation on the coast*'.

This does not imply that all ICZM projects are failing but it does suggest that the original overall expectations of achieving the 'conservation and sustainable multiple use of the coastal zone' are not being met. There are several reasons for this situation and these are related to a number of issues some of which involve the underlying root causes of environmental degradation in coastal areas. Key issues are:

1. *Failure to address the issue of population growth and overcrowding in the coastal zone.* The huge population pressure in many coastal areas makes it difficult to devise ways of achieving long-term sustainability. The much vaunted concept of 'sustainable cities', for example, is difficult to equate with the reality that major urban areas have huge footprints and that their inhabitants are largely unaware of where their food and energy comes from or how their waste is disposed of. The human footprint can be minimized but not eliminated; many management actions achieve a cleaner urban environment by moving the footprint somewhere else. Furthermore, current urban economies are largely predicated on the availability of cheap energy supplies and these are certainly unsustainable. It is difficult to see how sustainable coasts can be achieved without active measures (such as hugely improved education for men and women) to tackle population growth.
2. *Poverty and desperation.* With over 1.4 billion people living on US$1.25 or less a day (1995 figures), poverty continues to be a major factor governing people's

Figure 13.4 Aquaculture is an important coastal activity, albeit with environmental concerns that must be addressed. Pictured is a fish farm off Lamma Island, Hong Kong. © Hippo Studio/Shutterstock.com.

behaviour in many coastal areas. Halving poverty by 2015 is the first of the UN's Millennium Development Goals. The discourse of ICZM and of poverty reduction should be brought closer together for achieving sustainable coastal areas. Without tackling poverty, it will be difficult to deal with such issues as people living in coastal wetlands or fishing with explosives.

3. *Technocratic tunnel vision.* Scientific research is enormously valuable for supporting planning but scientists have often failed to recognize the context of their work. Some of us, for example, have witnessed the development of sophisticated and data hungry decision support systems that fail to engage with the realities faced by decision-makers themselves or are simply too expensive to maintain. There is also a blind faith held by many people that whatever our problems are likely to be, science and technology will come up with something to 'fix' them. There are still relatively few natural scientists who engage in

multidisciplinary research alongside social scientists and these two branches of science have their own mindsets and jargon. The outcome of this is a tendency to recommend 'end of pipe' or 'magic bullet' solutions to problems rather than to work out if the money might be better spent on dealing with their root causes. Huge expenditure in coastal sea defences, for example, might be saved by 'managed retreat' and investment in better planning and services for the displaced population or their activities.

4. *Denial of the reality of trade-offs*. Policy makers and development agencies appear to find it difficult to accept the uncomfortable reality that most planning decisions involve losers as well as winners. The 'win-win' solution (where nobody loses) has become a kind of Holy Grail of policy in many places but there are very few situations where it can be a reality. Unfortunately, this situation has led to projects being favoured that tackle what some people describe as 'the low hanging fruit'; things that can be dealt with easily in order to demonstrate success. Coastal fisheries management is an example of an issue where there are bound to be losers because much of the resource is already overexploited. Substitution of fishing with aquaculture does not necessarily help because much of the finfish aquaculture is fed by fish meal from wild fisheries somewhere else. Furthermore, an efficient aquaculture system can operate with considerable less people than wild fisheries so employment loss is inevitable. The debate needs to switch to the reality of trade-offs; understanding and quantifying them, minimizing the losers and finding mechanisms for them to be compensated.
5. *Predetermined views on consensus, democracy, culture and power*. There is view amongst some ICZM specialists that most problems can be resolved by coordination alone and that this can be arranged by consultation and consensus building. Whilst this may be true in the context of some Western-style liberal democracies, it cannot be generalized. People's worldviews are conditioned by their culture, upbringing and personal experience, and change very gradually through education. In the short term, it is necessary to understand the diversity of current political realities and work with them to the extent possible. There is a growing view that some of the more intractable management problems require strong governance, bearing in mind that some political decisions cannot be made from a populist standpoint. A decision to build an offshore wind farm or to create a marine protected area may be in the overriding public interest but may also provoke local discontent. Conversely, the development of intensive agriculture in river catchments such as the Danube or Mississippi may cause eutrophication and a 'dead zone' in the coastal waters of the Black Sea or Gulf of Mexico and therefore harm the economic and social interests of people in the coastal zone. All political systems have great difficulty in dealing with these issues. Governance

Figure 13.5 Some coastal developments take place with no environmental planning in evidence. Image by Laurence Mee.

has to be appropriate at all geographical scales but this is not easy to achieve in practice without clear norms and strong institutions.

6. *Issues of scale and the tangled web of globalization.* The problems facing the coastal zone are driven by social and economic processes across a wide range of scales. Within any one country, there are often tensions between central and local government. The central government, for example, may be driven by the need to reduce a national economic deficit through increasing the exploitation of its natural assets whereas local government may feel strongly about conserving them. Arguably, the geographical scale of these problems is increasing as a consequence of global market forces that are difficult to deal with because of the absence of effective governance. The development of aquaculture in shrimp ponds in Thailand is intrinsically linked to similar activities in Ecuador or Indonesia because prices are determined by the global market and the limited profit margin may constrain an investment in more sustainable practices. Trade, tourism, energy, agriculture, export fisheries and industry are all sectors highly sensitive to external drivers. It is often difficult to scale up successful ICZM projects to a regional and national level because each scale has different priority problems and ways to resolve them. An ICZM demonstration project in one coastal region may have successfully created sustainable tourism businesses as a way to achieve employment alongside the creation of a marine protected area. Successful replication along the coastline would result in competition for a niche market and be against the interests of the demonstration site.
7. *The legacy of the past and the long view of the future.* The development of a coastal area often 'locks in' social and ecological consequences that cannot be reversed on a decadal timescale. The excessive application of agrochemicals,

for example, can constrain groundwater use for decades or severely compromise any attempt at rehabilitation. The construction of a new housing development may change the nature of the coastal environment for centuries; few people would propose removing a badly sited development from the past even though it compromises future sustainability. Each development shifts the ecological baseline for the future and makes a return to the pristine past an impossible dream. The shifting baseline phenomenon is particularly difficult because human society has great difficulty in planning for intervals of more than a few years. Unfortunately, many projects funded in the name of ICZM have also failed to take the long view by creating new developments that may alleviate a social problem in the short term but compromise the long-term objective of sustainability.

The issues outlined above have led to considerable delays in implementing ICZM; local scale integration has been more successful than 'vertical' integration between sectors that operate at different geographical scales. Australia, for example, took a leading role in ICZM development almost 40 years ago but poor integration between State and Federal governments has, until recently, made these policies ineffective. South Africa has had difficulties because of the need to prioritize poverty alleviation and integrate the Sustainable Livelihoods approach with ICZM. In the European Union ICZM has suffered from a decision to limit policy to a relatively weak set of guidelines. None of this diminishes the important contribution of ICZM to coastal sustainability but the current pace of social and ecological change requires a 'gear shift' in policy development and one that facilitates a cross-scale approach.

Creating a new policy framework for the future

How can we build on the lessons learned from some 40 years of ICZM development? A number of other policy paradigms paralleled the development of ICZM. On the landward side of the coastal margin, Integrated Catchment Area Management (ICAM), a particularly useful approach for dealing with transboundary river basins, spurred the development of river basin management programmes. On the seaward side, Regional Seas programmes and the Large Marine Ecosystem concept provided geographical frameworks based upon pragmatic political boundaries and ecosystem boundaries respectively. Most of these management approaches faced similar governance challenges but remained quite poorly connected.

The UN Convention for Biological Diversity – another product of the Rio summit in 1992 – has provided an overall context by defining an 'ecosystem approach' to management in which humans were regarded as part of the ecosystem; 'people in nature' rather than 'people and nature'. This broader 'ecosystem approach' offers a

Figure 13.6 Solid waste disposal is a particular problem for small islands, and waste deposited in open landfills on hillsides often ends up in the sea. Image by Laurence Mee.

unified basis for achieving sustainability. There are various definitions of it but one of the simplest is: *A resource planning and management approach that recognizes the connections between land, air, water and all living things, including people, their activities and institutions.*

The Ecosystem Approach takes us back to our basic understanding of how coupled social–ecological systems are defined and operate. The CBD definition of an ecosystem is: *a complex of plant, animal and micro-organism communities and their non-living environment interacting as a functional unit.* It would be quite easy to extend this to include humans and their institutions but what would it really mean? In the past and in rural regions of our planet that we describe as 'developing', most humans were and remain part of ecologically defined functional units. Even in the so called developed countries, our grandparents generally ate local food that varied with the seasons. For those living in major cities and everywhere in the richer countries in the world, the situation has changed dramatically; food, fibre and energy are sourced from almost anywhere on the planet. This complete disarticulation of our consumption from our local environment makes it difficult to define system boundaries and to take full responsibility for our footprint; even solid waste is regularly transported across the planet.

Sustainability ultimately requires the management of a complex variety of interconnected systems and this requires entirely new patterns of cooperation that we have yet to devise.

In 1998, the CBD organized a major workshop in Lilongwe to move forward on defining a more practical basis for implementing the Ecosystem Approach and from this the following Malawi Principles emerged:

- Management objectives as societal choice
- Management decentralized and multi-sectoral
- Appropriate temporal and spatial scale
- Conservation of ecosystem function and resilience
- Appropriate balance between conservation and use
- Management within system limits
- The outward vision (respect interconnectedness) and long-term vision (change is inevitable)
- Broad use of knowledge, scientific and traditional
- Incorporation of economic considerations (costs and benefits, removal of externalities, etc.)

At first glance these principles do not look so different from the approach taken to ICZM. The important differences are the greater emphases on setting the right scale of time and space, interconnectedness between scales and the endeavour to conserve ecosystem resilience. Each problem may have its own temporal and spatial scales and these are not necessarily defined by physical geography.

One of the most significant elements of the Malawi Principles is the recognition that the future state of the natural environment will largely be determined by society and is thus an expression of human values. The human influence on the planet is so great that some scientists consider us to be living in a new geological epoch: the Anthropocene. This is an enormous social responsibility that we have been slow to grasp. Ultimately, it doesn't really matter how system boundaries are defined, provided that people are willing and able to take responsibility for their footprint at all levels, with knowledge of how system functions (including resilience) will be maintained and to exercise precaution where uncertainties are identified.

One practical means of achieving management in the face of uncertainty is the process of Adaptive Management. Originally developed in the 1970s by Charles (Buzz) Hollings, Adaptive Management has sometimes been paraphrased as 'learning by doing'. The relatively simple principle involves the following basic steps:

Figure 13.7 A busy coastal city, Benidorm, Spain. The huge population pressure in many coastal areas makes it difficult to devise ways of achieving long-term sustainability. © Philip Lange/Shutterstock.com.

1. A decision on the appropriate overall scale of the system based upon a mixture of scientific knowledge and political pragmatism.
2. A comprehensive assessment that includes information on the changing state of the natural and built system, the pressure on the environment causing the changes, the social and economic drivers of change (there are also natural drivers), the human welfare consequences of the natural system changes, and the institutional laws and policies available to manage the system. Foresight modelling or similar techniques can also be used to explore future scenarios for the system. All of this information should be presented in a way that is clearly understandable to all stakeholders and the general public.
3. A stakeholder decision on a desired *vision* for the future state of the system, perhaps at a point two decades into the future. This is an important decision and must be accompanied by objective success indicators. These should be regarded as achievable but there is no need to understand the exact pathway to achievement.
4. Agreement on the first step towards the vision. This should consist of 'hard' commitments to set and achieve *operational objectives* over a politically

Figure 13.8 The human footprint can be minimized but not eliminated; many management actions achieve a cleaner urban environment by moving the footprint somewhere else. © Artmann Witte/Shutterstock.com.

pragmatic timeframe (normally within a term of office of elected officials) using all appropriate legal, policy and technical instruments. Responsibilities for implementation must be defined; there must be full accountability and clear indicators of success.

5. Monitoring of long and short term indicators and complete public transparency of this information.
6. After completion of the first step, progress towards the long-term vision is measured and a new set of operational objectives are defined. As knowledge of the system grows and societal values change, it may be necessary to update the vision from time to time.

The value of this approach is that it defines a clear shared vision that is not stymied haggling on the details of exactly how it is to be achieved. It uses all available information and is highly pragmatic as it couples vision with political realities. Practical experience suggests that stakeholders with quite different sectoral perspectives will work together on long-term visions providing they are negotiated on the basis of how to meet their *needs* rather than their *positions*. There are two main risks however: (1) that the 'vision' could be a trade off between human aspirations and natural capital in a way that weakens ecosystem resilience, and (2) that adequate long-term monitoring systems are not established or are insufficient to permit the feed back and learning that allows the process to advance. There is also a risk that inherent conservatism in governments will block attempts to set ambitious long-term objectives or that they will want clear evidence-based plans for every step in the process before agreeing to them.

Adaptive management cycles can cascade across system scales. A vision for the system at a regional scale can be delivered by actions at the sub-regional scale, each following specific operational objectives. This is the case with the biggest current application of Adaptive Management: the European Marine Strategy Framework Directive (MSFD). The MSFD was adopted in 2008 and is the first EU-wide legislation for Europe's regional seas (the Baltic, north-west Atlantic, Mediterranean and Black Sea). Its primary objective is to reach 'Good Environmental Status' (GES) for each of these seas by 2020. GES is being defined for a number of sub-regions by the EU Member States through a complex process of fact-finding and consultation. GES is a measurable vision for Europe's seas for 2020. It will be achieved by coordinated individual actions by the EU Member States. These actions will be delivered through national scale policies and actions such as the new UK Marine Bill and the Scottish Marine Bill (Scotland has devolved responsibility for its territorial seas). In effect, the national actions will provide the operational objectives for the regional-scale vision. The national marine bills will also devolve some of their work to smaller geographical scales. At these scales, stakeholders will engage in marine spatial planning, a process that involves assigning rights and responsibilities and is informed by multidisciplinary science. This cascading from the EU to the sub-national has the advantage of combining 'top-down' (directed) and 'bottom-up' (consultative) management.

The MSFD interfaces with similar legislation for river basins and coastal regions: the Water Framework Directive (WFD). The WFD divides Europe's rivers and coasts into 'River Basin Districts' and sets a long-term obligation for each of these to achieve Good Ecological Status, a measurable vision statement supported by specific short-term actions. In addition, there are EU-wide guidelines for ICZM but these are not obligatory. There is relatively little enthusiasm by EU Member States to develop legislation and policy for ICZM; the general feeling is that by coupling the river basin and regional seas scales and requiring detailed contributory actions by Member States, there will be little need for ICZM.

Adaptive Management is also being tested on a global scale through the International Waters Focal Area of the Global Environment Facility. This has developed a process of assessment and strategic planning known as the TDA/SAP process (Transboundary Diagnostic Analysis/Strategic Action Programme). This also sets overall long-term goals and shorter term actions to attain them, coupled with environmental and process indicator monitoring to provide the necessary feedback. TDA/SAPs have been completed for systems in every continent but it will be some years before their effectiveness can be ascertained.

These wider integrated approaches are a logical development of the principles underlying ICZM rather than a departure from them. They allow better integration across scales and management of uncertainties but they do not explicitly deal with the underlying issues of population growth and poverty in coastal areas. These two fundamental concerns require a new social and economic order and a more serious global commitment to a sustainable future for humanity.

14 Hazards and warnings

DAVID PUGH

David Pugh was the founding chairman of the IOC Global Sea Level measuring system GLOSS. His scientific interests include tides, mean sea level and the statistics of coastal flooding. He is the author of *Tides, surges and mean sea level* (1996), which is being updated, and *Changing sea levels* (2004).

Introduction

The dream of many holidaymakers is to walk along the beaches that surround the Pacific Ocean. But, on 22 May 1960 this would have been very dangerous, and potentially fatal. More than 1600 people along the Pacific shores were killed by the tsunami generated off southern Chile by the biggest earthquake of the twentieth century. Ten years later, on 12 November 1970, an extreme cyclone generated a 9-metre storm surge which flooded the low-lying islands of the Ganges Delta, causing more than 300 000 deaths.

The United Nations defined the 1990s as the International Decade of Natural Hazard Reduction. Hazards include tsunami as well as earthquakes, volcanic eruptions, storms and landslides. The sea has its own specific hazards; besides the coastal flooding impacts discussed here, those affecting marine living resources, human health and the safety of marine shipping are discussed elsewhere in this volume.

While many of these hazards cannot be avoided they can be anticipated, and preparations made for appropriate life-saving responses. Economically, prevention or at least mitigation of the impact is preferable to expensive and extensive relief and construction operations in the wake of disasters.

There have been several serious marine flooding events in the past 50 years from tsunami and storm surges; each has stimulated some partial local response.

Figure 14.1 The GLOSS sea level gauge on Easter Island. The solar power panels and the satellite transmission aerial are visible. Image by David Pugh.

Four years after the end of the International Decade of Natural Hazard Reduction, the Indian Ocean tsunami struck: and there was no warning. Over the past 50 years there have been major improvements in our preparedness, but there is still an innate apathy among the public, and therefore among politicians, to give priority to events that may occur only once in a lifetime or longer. Hospitals and bridges make a more immediate demand on the public purse, though even modest investment in warning and response systems for disasters is very cost-effective.

The oceans can be very hostile neighbours. The theme of this chapter is the need for a coordinated global approach, perhaps a Marine Hazards Convention, to anticipating and mitigating the effects of all marine hazards. Without sustained commitments, we cannot be fully prepared for the inevitable. Yet governments are cautious about relying on truly international systems, understandably favouring parallel local systems under their direct control. Developments are spasmodic, often triggered by the latest disaster

People, especially those living near the sea, are rightly asking what measures their governments are taking to provide warnings of future events so that impacts can be mitigated. In turn, governments are asking scientists to provide them with early warning systems. Critical questions include: how much warning time is possible; will warnings be reliable; and, of course, how much will it cost?

This chapter looks at the motivation for and operation of some of the warning systems developed by governments in response to coastal flooding disasters.

Tsunami

Our ability to give hazard warnings depends on the physics of the triggering events, and on the way in which the subsequent ocean waves propagate.

Figure 14.2 Easter Island statues reinstalled after the 1960 tsunami. The statue on its back in the foreground, which weighs several tonnes, was carried inland 250 m; it has been left in place as a reminder of the power of that event. All Easter Island stone statues are installed with their backs to the sea. Image by David Pugh.

For tsunami, the usual trigger is a submarine earthquake, though meteorites, landslides into the sea and slumping of sediments on the continental slopes are also possible causes. Not all submarine earthquakes produce tsunami. The important element is a vertical movement of the crust that displaces the seabed and causes a tsunami to be generated in the water. The tsunami will then propagate with a magnitude that depends on the amplitude and extent of the seabed displacement and at a speed in the deep ocean given very accurately by the formula:

$$\text{Wave speed} = (\text{gravity} \times \text{water depth})^{1/2}$$

Therefore, using the above formula, and for a typical ocean depth of 4000 m, the speed is 715 km per hour, similar to the speed of an intercontinental jet plane. The wave from the 1960 earthquake in Chile took 15 hours to reach Hawaii, and 22 hours to reach Japan. For people living near the source in Chile, the delay before the tsunami arrived was only a few minutes, and the height of the wave more than 25 m.

The tsunami propagation speed is used to calculate the travel times of the waves to different coastal cities and shores, which can then serve as the basis for tsunami warnings: the greater the distance from the source, the longer the possible warning time, and usually, because the waves weaken as they spread out, the smaller the impact. In the ocean the tsunami may have an amplitude of a metre or less, and may pass undetected by ships at sea, because of the very slow rise and fall and the presence of the much more evident surface waves. As the waves approach shallower water the speed decreases and the size can increase dramatically. Along coastlines tsunami can trigger local sea oscillations, called seiches, with periods of only a few minutes, which persist for hours, or even days. These local seiches are the main cause of the damage and devastation.

Figure 14.3 Marine flooding makes access difficult to these prestigious water-front apartments in Cartagena, Columbia. A meteorological surge and a high tide coincide; it happens from two to three times a year. Image courtesy of Julián Reyna.

Seismologists cannot yet tell us when or where the next earthquake will occur, but they can say where major seismic events are most likely. We know that the seismically active plate boundaries of the Pacific Rim are prolific generators of earthquakes, with the attendant risk of tsunami. Earthquakes are less frequent or intense elsewhere; but the threat of tsunami exists for the Atlantic Ocean and, of course, for the Indian Ocean.

A tsunami warning system has three vital components. The first is a seismic network to detect and locate submarine earthquakes or other triggering events. The second is a system for measuring sea levels close to the source to see whether a wave has been generated, as in many cases there will be no wave, and so no tsunami. As with all warning systems, what may appear as false warnings must be minimized, as too many will make people indifferent, so they later ignore the warning that precedes a genuine disaster. The final element of a warning procedure is the local system of issuing warnings, and organizing mitigation measures, such as evacuation to higher ground or the closure of flood defences.

Figure 14.4 In the foreground is the sea level station on Christmas Island, Indian Ocean. © National Tidal Centre, Australian Bureau of Meteorology.

The first two components, detection and forecasting the propagation and arrival of a tsunami wave, are dependent on international cooperation. The organization of appropriate coastal responses is of course a matter for national and local authorities.

Five years after the 1960 Chilean earthquake (its magnitude of 9.5 on the Richter scale made it the biggest of the twentieth century), governments set up an ocean-wide warning system. In 1965, the United Nations Educational, Scientific and Cultural Organization's (UNESCO) Intergovernmental Oceanographic Commission (IOC) accepted an offer made by the USA to expand its existing Tsunami Warning Centre in Honolulu to become the headquarters of an International Pacific Tsunami Warning System and at the same time invited other IOC member countries to integrate their existing facilities and communications into this System. A meeting was held in Honolulu, Hawaii in 1965 establishing the International Tsunami Information Centre (ITIC) and an International Coordination Group for the Tsunami Warning System (ICG/ITSU). The System has operated continuously since then. Today, 26 IOC Member States around the Pacific Ocean participate, operating a network of seismic and sea level gauges, and a rapid communication of information, to and through the Centre, located in Hawaii. The Information Centre covers tsunami awareness and the application of best practice for adapting to tsunami risks.

Operationally this System was restricted to the Pacific Ocean, although there is a well-recorded history of tsunami occurring in other oceans. These include, for example, the Lisbon earthquake and tsunami of 1755 and the Krakatoa event of 1883. In the absence of any recent major tsunami events there was never enough political will or priority to set up more global warning systems, until the devastating, recent (2004) event in the Indian Ocean.

Political priorities changed dramatically immediately after the Indian Ocean tsunami generated by an earthquake measuring 9.2 on the Richter scale off Indonesia on 26 December 2004. In the days and weeks immediately after the tsunami, regional governments were keen to establish national warning systems. Because of its Pacific experience, the Intergovernmental Oceanographic Commission was asked to lead intergovernmental cooperation within the United Nations organization. Local enthusiasm was quickly channelled into an ocean-wide system, based on the existing system in the Pacific, by Patricio Bernal, the Executive Secretary and his colleagues. The Indian Ocean system formally became operational in June 2006.

Post-2004 the immediate priority was to establish the new Indian Ocean Tsunami Warning System. However, the goal is also to develop systems in other ocean areas, leading eventually to a truly global system.

To provide this wider coverage for tsunami warnings, Intergovernmental Coordination Groups (ICG) have recently been established for the Caribbean, the Atlantic and the Mediterranean, under the aegis of the IOC, in cooperation with the Pacific Tsunami Warning Centre (PTWC) from the USA and the Japan Meteorological Agency (JMA).

In theory these two centres can provide a global warning service for all tsunami. Information from globally distributed seismic and sea level gauges can be transmitted rapidly and accurately, provided of course that communication systems are not themselves damaged by an event. Some duplication is necessary as a safeguard. Nevertheless, there are good reasons, both technical and political, for governments to establish their own local warning centres to enhance the forecast of local impacts and issue warnings in a way that a global centre cannot do. One of the key elements is the local buy-in to global systems. Local advanced planning for education and awareness of the dangers and of appropriate local immediate actions once a warning has been received is essential.

Surges

Tsunami are caused by single rapid events; however, winds and air pressure, acting over hours and days, can also generate extreme sea levels. Coastal flooding due to extreme weather effects can have devastating consequences. Thousands of lives have been lost and damage of many billions of dollars can be attributed to these events. The effects of tropical storms and extra-tropical storms on sea levels are different. A clear distinction is usually made between them in terms of flood warning generation. The length and space scales of the impacts along the coast are also very different.

Figure 14.5 High tides and storm waves damage important coastal infrastructure, including railways. © Four Oaks/Shutterstock.com.

There are two ways in which the weather affects sea levels. First, changes in atmospheric pressure will depress or elevate the sea surface, which in extreme cases can generate dangerous surges for coastal populations. Second, forces due to wind drag at the surface can also result in coastal inundations with heights determined by several factors such as wind strength, direction and the length of time for which they act. Weather prediction is becoming very sophisticated and many countries have their own computer models and warning systems in place. However, these systems depend upon the intergovernmental cooperation in regional and global monitoring and data exchange, elements of which have been developed over the past century and a half. Even now the science is not exact and scientists continue to cooperate to improve the predictive skills of the computer models. For the less fortunate countries, without the necessary skills and technology, regional cooperation and intergovernmental assistance is essential.

Figure 14.6 Children returning from school through floodwaters during a tidal surge in Chittagong, Bangladesh. © Jashim Salam/Marine Photobank.

Extra-tropical storms, found in mid-latitudes, extend over hundreds of kilometres around the central region of low atmospheric pressure and are usually relatively slow moving. They affect large areas of coasts over periods that may extend to several days. The various local physical processes require regional cooperation in the generation of warnings. For surge warnings the first essential is good weather forecasting; this is why the national meteorological authorities operate many of the surge warning systems. In the Baltic Sea, extreme weather raised sea levels by 4 m in St. Petersburg in 1924; earlier serious events occurred in 1777 and in 1824. Defences for St. Petersburg are now being built across the River Neva.

The North Sea, between Britain and northern Europe, has been described as a 'splendid sea for storm surges'. It is open to the North Atlantic Ocean in the north so that the extra-tropical storms, which often travel across this entrance from west to east, are able to set the water in motion with very little resistance from bottom friction. Surges are generated by winds acting over the shelf to the north and north-west of Scotland, and by pressure gradients travelling from the deep Atlantic to shallow shelf waters. When these water movements

Figure 14.7 Venice has adapted to rising sea levels. Image by David Pugh.

propagate, from the north, into the North Sea they travel down the east coast of the UK before impacting the western European coast. The progression takes several hours, which allows time for flood warnings to be issued farther along the coast. Cooperation between the regional governments gives warnings in time for sea defences to be activated. Fifty years ago this information was transmitted by post office telegrams or later by telephone. Now we have the Internet.

Tropical meteorological storms are seasonal and small in scale but very intense. Their direction and associated landfall is still relatively unpredictable. They produce exceptionally high flood levels within a confined coastal region of perhaps tens of kilometres. But only a few kilometres away the winds from the same storm are directed offshore, giving very low sea levels. Tropical storms are known variously as hurricanes (USA), cyclones (India), typhoons (Japan), willi-willies (Australia) and baguios (Philippines).

As an example, around the Gulf of Mexico and along the Atlantic coast from Florida to Cape Hatteras the greatest risk of flooding comes from tropical storms – hurricanes – that originate in the tropical Atlantic Ocean from where they travel in a westerly direction until they reach the West Indies. Here many of them turn northwards towards the coast of the USA. Their greatest effects on sea level are confined to within a few tens of kilometres of the point at which they hit the coast.

In 1969, Hurricane Camille raised levels by up to 7 m on the Mississippi coast, and more recently Hurricane Andrew in 1992 raised sea levels locally in Miami by 5.1 m.

The coasts of India and Bangladesh that surround the Bay of Bengal are very vulnerable to severe flooding due to tropical cyclones, with terrible consequences as referred to above. Several thousand people were killed in September 1959 when Typhoon Vera struck Japan, producing a peak surge of 3.6 m at Nagoya on the south coast of Honshu. Other vulnerable regions include the coasts of southern China and Hong Kong, the Philippines, Indonesia, Northern Australia and the Queensland coast. In all cases the surge levels and the damage caused by the storms are very sensitive to the direction and speed of their progress, so that no two storms have exactly the same effects. In many cases flooding risks have been increased in recent years by urban developments, roads and concrete surfaces, which increase the immediate outflow from rivers and are sometimes coincident with surges from the sea, generated by the same extreme weather events.

The regional nature of extreme weather means that local arrangements are needed to generate effective flood warnings. Effective warnings are dependent on the application of numerical models of atmospheric movements, and their effects on local sea levels. The World Meteorological Organization assists in developing the local infrastructure, and in the development of appropriate computer simulation systems.

Mean sea level

Tsunami and surge flooding produce extreme sea levels, relative to the average, or mean, sea level. As mean sea levels continue to rise through the twenty-first century, these increases will add to the total extreme levels that impact the coasts. The increasing frequency with which the city of Venice, at the head of Adriatic Sea, is subject to flooding by high sea levels has attracted much attention in recent years. These flooding events, generated by weather over the Adriatic, happen more often than in historical times because of a gradual increase of mean sea level, relative to the land, of between 3 and 5 mm per year.

Globally the mean level of the sea has increased by 15 to 20 cm in the twentieth century. There are significant local variations, due especially to the processes of recovery from the last ice age. As a consequence of the unrepresentative distribution of long-term sea level measurements over the twentieth century, and the uncertainty in the correction for land movements, a true global twentieth century mean sea level increase can only be estimated. Measurements of sea level from globally orbiting satellites will be the basis for identifying mean sea level changes in the twenty-first century. Satellite altimetres can now measure globally averaged sea levels with resolutions of better than a centimetre.

Figure 14.8 Damage after the 2004 tsunami at Ache, Indonesia. © A. S. Zain/Shutterstock.com.

Projections for a warmer 'greenhouse' world suggest increases over the next century of 50 cm or more. These slow, sustained increases in mean sea level can be anticipated based on detailed scientific studies, though the risks of rapid collapse of grounded ice sheets remains a major uncertainty in future estimates. As explained above, the mean sea level increases must be added to the risks of flood levels already known from storm surges. Scientists are also looking at the risk that the intensity and frequency of storms and surge flooding will increase in the future. People often ask whether the rate of sea level rise is accelerating. Against the background of natural shorter-term variability it has proved very difficult to detect any changes in the *rates* of mean sea level rise, although these are anticipated in theoretical responses to global warming.

A sensation-hungry press often dramatizes possible mean sea level rise by showing maps of changed coastlines for rises of 5 m or more. The expected change over the next hundred years of around 0.5 m may be an order of magnitude less, and perhaps the potential impact will be much more modest; but it is still a serious issue unless preventative measures are planned and implemented. In all discussions of the impact of an increase in mean sea level, we must remember that the *rate* of change is equally as important as the *magnitude*. If the rate is slow enough, then natural, social and

Figure 14.9 Melting of mid-latitude glaciers has contributed to sea level rise in the past 100 years. © Ribeiro Antonio/Shutterstock.com.

economic systems will adapt at their own rates. Politicians under pressure through the alarmist reports of the press need the defence of systematic scientific studies as a basis for rational and progressive planning.

Mean sea level has become and remains an active area of research, stimulated both by popular concern about global warming and by a range of new measuring capabilities such as satellites and GPS. The Intergovernmental Panel on Climate Change regularly reviews and updates the evidence. Their fifth report is planned for 2014.

Other

Coastal flooding due to sustained waves and the swells from distant storms can cause damage, especially where the tidal ranges are low, as for tropical coral islands. Serious flooding due to the sustained swell of distant storms is a regular occurrence in the Maldive Islands, but there is no systematic warning procedure.

As beach visitor numbers increase globally, understanding the physical hazards and risks posed to the beach user within a worldwide context becomes even more important, and when properly communicated can increase the confidence and hence the numbers of people enjoying these facilities. Surfers enjoy riding the largest waves, but these are also a hazard for normal bathers. Reliable warning systems can allow both groups to plan ahead.

Conclusions

This chapter has looked at the basic physics of coastal flooding, and shown how the development of warning systems depends on the local conditions. These natural extreme events are not controllable but they can be anticipated, both as long-term

Figure 14.10 This dyke in the India Ganges river delta is under threat from sea erosion, and rising sea level. © Joerg Boethling/Still Pictures.

changes in flood statistics for design purposes, and for the issuing of immediate warnings.

Most extreme environmental events cannot be avoided: their power is way beyond the control of humans. But they can be taken into account in planning coastal structures, and lives can be saved through education and adequate warnings, before the full effects of the waves and storms arrive. The social and economic costs can be huge. The 2004 tsunami havoc could add up to an overall capital loss of US$15 billion in India, Indonesia, Sri Lanka and Thailand.

Tsunami need a global system of warnings based on tide gauges and seismographs, and efficient local response mechanisms to educate and in emergency situations, warn the public. Storm surge warning systems rely on good regional weather forecasts and detailed computer simulation of their progress; they are more local in their impact. Monitoring global mean sea level changes through the twenty-first century will depend on sustained long-term global measurement programmes.

Warning systems must be based on good science. Although for imminent events, exact forecasts are not as essential as is the timely delivery of the warnings, there are perils in making too many imprecise flood warnings. Not to issue a warning of a disaster would be a serious error. However, too many false warnings lead to public complacency, and may lead to future warnings being ignored, with equally serious consequences. The issuing authorities need the best possible forecasts to avoid this confusion. Fortunately technology and scientific understanding have made great progress in the past 50 years. So too have the methods of communication, meaning that very few communities are now isolated. A warning issued by one country will soon be broadcast to its neighbours and worldwide.

Despite the effective technical capabilities of regional and global coordination, the responses from governments are still piecemeal and partial. While the Indian Ocean now has a tsunami warning system, similar systems in the Atlantic Ocean are slow to develop. Inevitably politicians, while responding immediately after major disasters, soon find competing claims on their attention and resources. In order to show their electorate that they are effective and in control, politicians favour funding national systems rather than support the establishment of global networks. The situation has improved over the past 50 years, but much remains to be done, and global systems remain critically under funded.

There are many UN bodies with a partial responsibility in this area, in particular the WMO, the IHO, UNEP and the IOC. The Intergovernmental Oceanographic Commission of UNESCO, inspired by the 2004 tsunami, has stated the requirements for international cooperation in a 2005 Resolution. A global tsunami and other ocean-related hazards early warning system should:

- build on existing mechanisms and capabilities and should address all the necessary components for an integrated end-to-end system, which are hazard and risk assessment, warnings and preparedness;
- recognize the IOC's comparative advantage in ocean-related observations, data management, forecasting, dissemination of forecasts and warnings, and capacity building;
- improve our understanding and forecasting of ocean hazards to reduce their impact through focused scientific research and technological development;
- work to improve durable global observational capabilities, particularly through the transfer of marine technology; and
- make the information available to the entire world.

The local response mechanisms are the most important part of all warning systems, but sustained global systems to inform them are also essential. The ability of governments to act coherently in the development of global marine flood warning systems would be strengthened if there were a Convention on Marine Hazards, relating to the exchange of expertise, data and resources, and within which comprehensive, rational and progressive systems could develop.

15 Regional cooperation: the Caribbean experience

GUILLERMO GARCIA MONTERO

Guillermo Garcia Montero is Director of the National Aquarium of Cuba. His scientific work is mainly devoted to coastal zone management, international cooperation and capacity building. He is Chair of the Cuban National Oceanographic Committee and has been a member of several IOC Bodies. He was awarded the Order 'Carlos J. Finlay', Cuba's Council of State's highest national recognition of those working in science.

Introduction

Not surprisingly, many important intergovernmental arrangements in ocean science focus on regional cooperation. Like the terrestrial environment, with its geographical differences of mountains, deserts and plains, ocean regions can be quite distinctive and regional governments can cooperate effectively to tackle mutual issues and decide on common priorities. It would require several volumes to try to describe all regional marine arrangements around the globe, but the Caribbean and the IOCARIBE provides an excellent example.

IOCARIBE

IOCARIBE has been, and continues to be, one of the most important regional programmes of the Intergovernmental Oceanographic Commission of UNESCO. Its mandate to bring together, and to organize, the international marine scientific community at regional scale, provided a strategy for the promotion, coordination and cooperation in the important and diverse Caribbean region.

With the single exception of the USA, most Caribbean States depend largely on their marine and coastal natural resources. The Caribbean is made up of a great diversity of states and cultures, and distinct geographical characteristics. Far from

Figure 15.1 Participants at the 2005 tsunami meeting in Mexico. © IOC/UNESCO.

dividing our countries, the sea constitutes a common bond that unites all nations. The regional Member States collectively need to meet the challenge of sustainable development to ensure their future.

In the opinion of the author, the basis of the efforts of IOCARIBE for the past 40 years has been to increase the political will of individuals, institutions and countries to cooperate for the common good of science and the well-being of the people, to bring the marine and coastal science and related professionals closer together to improve resource management and to promote science as a force for sustainable development.

The strategic objectives of IOCARIBE have addressed the needs for capacity development, and the use of those resources to develop marine and coastal scientific and services programmes of high priority for the region's countries. In its proud history, there have been many successes, advances and, of course, setbacks.

The Growth of a Regional Organization

Since the foundation of the Intergovernmental Oceanographic Commission in 1960, the '*international groups for coordination of the cooperative investigations*' remain the most important subsidiary bodies for international scientific activities coordination. IOCARIBE began in 1968 with an innovative research initiative, the 'Cooperative Investigations in the Caribbean Sea and Adjacent Regions' (CICAR), a programme to which it owes a great deal of its current structure. In 1975, an experimental regional association was formed, which in turn led to the present regional Sub-commission. IOCARIBE is today a mature subsidiary regional body of the IOC and a leading regional organization in the field of international marine scientific research and cooperation. Its evolution through these three stages provides an interesting history of regional development.

Figure 15.2 Fishing in Belize. Ocean resources are an important part of Caribbean life. © Sharon K. Andrews/Shutterstock.com.

CICAR (1968–1975)

Let us begin with this first regional body. CICAR was the result of a decision by the IOC Assembly in 1967 to establish it as an IOC official programme. It was the result of a proposal made by the delegation of Holland to develop a 'cooperative' effort of scientific investigations in the Caribbean Sea and the adjacent regions.

Subsequently, the First Scientific Symposium and Meeting of the International Coordination Committee of CICAR was held in 1968 in Curacao, Dutch Antilles. This pioneer meeting had the goal to review the state of knowledge on the Caribbean marine environment and prepare a plan for 'future' cooperative efforts. The meeting was attended by delegates from Germany, Cuba, the USA, France, Jamaica, Mexico, the Netherlands, the UK, Venezuela and observers from Argentina, Brazil, Colombia, Trinidad and Tobago, and Uruguay. This was the first example of a marine international cooperation in scientific research and coordination in the Wider Caribbean under the auspices of an international organization. CICAR also provided the first and necessary step of characterizing the conditions of a rich marine region and determining the state of its marine resources and available scientific knowledge.

Conceptually, CICAR was originally structured as a scientific programme but proved to be a driving force in the development and evolution of intergovernmental cooperation in the region. In particular, the programme highlighted the shortage of scientific capacity and the lack of adequate research infrastructure in the region as a severe problem preventing the effective participation of many of the regional Member States. The development of marine scientific researchers and technicians in developing countries was recognized as a priority and the first regional meeting of the

Figure 15.3 Capacity Building and regional cooperation are fundamental to the Caribbean ocean science community. © IOC/UNESCO.

IOC Working Committee on TEMA (Training, Education and Mutual Assistance) was held in conjunction with the Seventh CICAR meeting and was specifically oriented to the needs of the wider Caribbean. CICAR successfully completed its 6-year plan with a symposium and scientific report containing several reports of results and databases and a set of important recommendations for future work. At the same time, the Member States recognized the need for a coordination mechanism that could build on the success of CICAR and recommended '... *the IOC establish a new subsidiary body for marine sciences cooperation in the Caribbean and Adjacent Regions*'.

The Association (1975–1980)

In responding to the request of CICAR, the 'IOC Association for the Caribbean and Adjacent Regions' was established by the IOC Assembly, in 1975. The Resolution '... *decides to create, with an experimental character, and for a six year period, an IOC Association for the Caribbean and Adjacent Regions, with the purpose to continue and develop the regional cooperation in marine sciences* ...'

The founding members were 13 states from the region (Colombia, Costa Rica, Cuba, Dominican Republic, France, Guatemala, Haiti, Jamaica, Mexico, Panama, Trinidad and Tobago, the USA and Venezuela) and three external, but equally active, states (Brazil, the Netherlands and the USSR).

Following CICAR recommendations, the Association established the general objectives of the intergovernmental mechanism, which would be later called IOCARIBE, namely to:

Figure 15.4 A priority is to sustain the fish habitat and the biodiversity of the Caribbean region. The coral reefs are particularly vulnerable to the immediate impacts of tourism and the longer term effect of global warming. © Yang Xiaofeng/Shutterstock.com.

1. increase marine scientific and technological development;
2. broaden the knowledge and optimal use of their natural resources;
3. unify resources for the solution of common marine problems; and
4. identify and define regional marine problems or those of common interest for two or more countries, settling down and coordinating measures for their solution.

IOCARIBE and CICAR were clearly different. When the IOC Secretary addressed the first meeting of IOCARIBE in Caracas, 1976, he underlined '... *IOCARIBE is not a continuation of CICAR with a different name, but a completely different body*', emphasizing the fact that IOCARIBE had been established in the region with the purpose to promote and stimulate the broadly recognized benefits of international cooperation and mutual assistance for marine sciences. Without a doubt, countries in the region wanted to continue the science programmes pertinent to their needs

Figure 15.5 Sediment loads entering the Caribbean Sea off the Meso-American coast. © Malik Naumann/Marine Photobank.

and priorities on marine resources, coastal management, marine pollution and on the urgent national needs for capacity development. IOCARIBE undoubtedly constituted a 'regional association' in a practical sense.

A main issue for the IOCARIBE Association was the lack of adequate finances. It is an issue that was also experienced by CICAR and one that continues to this day. Regrettably, in the development of many regional subsidiary bodies, Member States fail to understand the financial commitment necessary to fulfil the organization programmes and goals. However, the evaluation of the IOCARIBE Association made in 1982 pointed out that '… *the negative elements have been exceeded by the positive development of their programs and activities, even if several problems impeded the desired and foreseen development of the Association*'. Other deficiencies included a lack of participation by many of the regional Member States, poor contact between the Member States and the Regional Secretary, and a shortage of financial support for programmes, activities and the Secretariat function.

Independently of its contribution to marine sciences in the region, the IOCARIBE Association's most significant achievement was to provide a systematic mechanism of communication, serving as an international forum for exchanging of experiences and dialogue among its member states. In spite of the problems and obstacles during its 6-year history, the Member States of the Association decided to propose to IOC that IOCARIBE should be continued and evolved beyond its pilot phase as a permanent subsidiary body of the Commission.

IOCARIBE: SUB-COMMISSION (1982…)

After the 6-year experimental period as an 'Association', the Member States, gathered in Cancun in December 1980, recommended the creation of a new and superior mechanism for the promotion and coordination of international cooperation. The IOC Assembly in 1982 responded by establishing the IOCARIBE Sub-commission, the first of its kind.

According to the Regional Subsidiary Bodies' Terms of Reference defined by IOC, IOCARIBE as a Sub-commission will '… *promote international cooperation and coordinate research programmes, services and capacity building in order to learn more about the nature and resources of the ocean and coastal areas and to apply this knowledge for the improvement of management, sustainable development and the decision-making process of its Member States. The Sub-Commission is responsible for the promotion, development and coordination of the IOC's global scientific and research programs and ocean services…*' in the Caribbean and Adjacent Regions. The first IOCARIBE Sub-commission meeting was held in 1984 in Curacao, Dutch Antilles. At the time of writing a total of 10 Sub-commission meetings have been held.

An evaluation of the work of the Sub-commission during its first 10 years recognized that weaknesses existed, but overall the report was positive. In the author's opinion, more than 10 years after this evaluation, the balance continues to be positive. The Sub-commission continues to learn from its errors and profit from its mistakes.

The stated purpose of IOCARIBE is '*to promote international cooperation and coordinate research programmes, services and capacity building in the area of the Caribbean Sea, the Gulf of Mexico and Adjacent Regions, including the Antilles, Central America, and the north coast of Brazil, in order to learn more about the nature and resources of the ocean and coastal areas and to apply this knowledge for the improvement of management, sustainable development and the decision-making process of its Member States*'.

IOCARIBE, when establishing its programmes, keeps in mind the specific interests and necessities of its member states in the region.

IOCARIBE can be considered an international system created for coordination and promotion of sciences and operational services in marine and coastal issues associated with the region. Its main objectives are to generate and share knowledge, to assist its member states in capacity building, to promote and broaden cooperation with other international and intergovernmental organizations in the region and to provide regional liaison with global ocean programmes.

Figure 15.6 Tourism is of major economic importance to the Caribbean, but its environmental consequences are a concern. © Graça Victoria/Shutterstock.com.

Within the Caribbean, IOCARIBE develops programmes and projects of strategic importance at national, regional and global scales. These activities are the cornerstone of its existence; encouraging joint efforts, developing coordination and promoting cooperation.

The more notable current programmes are:

- **Sustainable Management of the Shared Living Marine Resources in the Caribbean Large Marine Ecosystem**, an important GEF project supported by 25 Member States in the Region, as well as Representatives from the Implementing (UNDP) and Executing Agencies (IOC-IOCARIBE) and other UN agencies such as FAO and UNEP, and NGOs such as The Nature Conservancy.
- **Harmful Algal Blooms in the Caribbean**, a programme established to create and improve research and monitoring capacity on algal blooms in order to mitigate the toxic effects of these events in the region.
- **Tsunami and Other Coastal Hazards Early Warning System in the Caribbean Region**. The Caribbean region is frequented by hurricanes and other marine hazards. In addition evidence exists of rarer, but potentially dangerous, occurrences of regional tsunami. IOCARIBE is working in cooperation with other partners to establish a regional integrated and sustainable multipurpose system to answer the needs of the Region in observation, warning and mitigation of these coastal threats.
- **ODINCARSA: Oceanographic Information and Data Network in the IOCARIBE and South America Regions**. Since its establishment in 2001, this programme has provided a cooperative network for the proper management and exchange of oceanographic information and data in the region.
- **Marine Pollution Research and Monitoring (CARIPOL).** This programme dates back to the times of CICAR and its success generated the CEPPOL Programme for

Research, Monitoring, Control and Reduction of Marine Pollution in the Caribbean in cooperation with IOC and UNEP. Its main goal is to find solutions to problems related to land-based pollution in support of integrated coastal zone management.

- **Capacity Development in Marine Sciences, Services and Ocean Observations.** 'Capacity development' remains a central objective of all programmes and projects of the Sub-commission. Recently, with the support of the IOC Unit for Capacity Development, IOCARIBE has begun a specific programme to reinforce the capacities of leadership in the research institutes of the Region and to reinforce the networks of cooperation among the marine sciences institutes in Member States. A related programme with UNEP and with the financial support of the Swedish Government is the *Caribbean Regional Network in Marine Science and Technology* established to enhance the capacity of the countries of the Wider Caribbean Region to address land-based sources of pollution and to improve the knowledge about the status and quality of the marine environment and its resources.
- Regional cooperative programmes exist to promote the development of marine charts, sea-level gauges and other ocean observations necessary for the safe and effective management of marine and coastal activities in the region.
- A demonstration of the maturity of the organization in the international arena is provided by the 'IOCARIBE AWARD'. This award, established in 1995, is presented at each Regional Assembly to those that have contributed significantly to the development of marine sciences in the region.

Main lessons learned

As pointed out above, successes and failures, tests and errors, and advances and setbacks have all contributed to the concept, design and development of what is today the Sub-commission. Most are perhaps very logical, but unfortunately it has not always been possible to address them or take them into account. The following lessons should be noted:

1. The negative impact of the severe gap between the needs of the regional marine and coastal sciences and services and the resources available.

   The lack of human and financial resources is the single most important limitation on the development of regional cooperative programmes. This is not a recent problem, but a continuing and chronic issue. It is also a problem that has failed to be addressed. In the author's opinion, this is perhaps due to the failure to establish regional priorities, which leads to the second lesson learned:

Figure 15.7 Regional weather-related events can be very serious. View of a house after a hurricane passed by in Freeport on Grand Bahama Island. © Ramunas Bruzas/ Shutterstock.com.

2. The positive result obtained when high-priority attention is given to the most urgent national needs.

   The Caribbean is characterized by economic and deep cultural roots associated with the sea and coasts. As a result, the most urgent national needs for Member States of the Sub-commission in the three main strategic fields of tourism, trade and transportation are, without exception, related to the marine and coastal environment. Bearing this in mind, the regional priorities have always been focused on the following main issues:

   - tourism
   - fisheries
   - marine contamination
   - coastal populations
   - marine biological diversity
   - marine security
   - storm surges
   - weather forecast
   - hurricanes, tsunami and other coastal threats.

3. Avoid automatic regionalization of global programmes … always look first to Regional needs.

   IOCARIBE has provided a catalyst for national programmes and a stimulus for marine sciences development in the region. But, towards the end of the 1980s and 1990s, regional efforts began to lose importance compared to global priorities, as reflected by most international organizations, including those of the IOC. As a consequence of this, training, education and other scientific activities were diverted to global interests that, although very important, are far from the most urgent needs of the regional Member States.

   The high priority accorded to global programmes and activities negatively influenced the development and overall performance of regional subsidiary bodies including IOCARIBE. To be fair, other central factors exist that influence this situation, for example:

   - worsening of the economic crisis of the region;
   - insufficient capacity to obtain support from external funding sources;
   - ineffective integration with other organizations and/or mechanisms with similar interests in the Region; and
   - insufficient national representations in meetings and other intergovernmental activities.

   In the author's opinion, the essential problem of this period was the erroneous and almost automatic application of the precept: 'Think globally and act locally'. This precept demonstrates the opposite to what many have practised. That is to say, it should not imply automatically the implementation of global programmes at regional and national scales. Cooperating to address and find solutions to the problems of each region or country is the only way to contribute to the solution of larger scale problems. It is also necessary to remember that a response to national and regional development will have a positive net impact in the international performance of Member States.
4. Recognize and develop institutional and personal leadership in all activities and programmes.

   The most successful programmes of the Sub-commission have been the result of a combination of several factors. However, institutional and personal leadership is the one that constitutes the core of success. To be successful coordinated activities need inspiration, direction and authority. Leadership is essential. Without leadership there can be neither integration nor coordination.

   The above-mentioned is not achievable if the necessary political will and support doesn't exist; therefore, it is essential to:
5. Catalyze, develop and/or increase the political will of individuals, institutions and countries, to address the essential problems of regional marine and coastal environments.

Figure 15.8 The US Virgin Islands, one of the world's most beautiful coastal environments. © Ken Brown/Shutterstock.com.

'Political will' should emerge from a conscious plan of oriented actions that researchers and science leaders need to develop in order to influence politicians and decision-makers and to generate actual change. In this regard, the science community has an enormous duty. It must demonstrate to policy makers and to the general public the effectiveness of scientific results, data and information in planning, preparing and implementing wise decisions for real resources management and sustainability. In addition, scientific results must be understandable; the information they generate has to be interpreted and accepted by a non-scientific community in order for it to be transformed into actions.

IOCARIBE and capacity development

The establishment of IOCARIBE was an unquestionable and strategic means for the promotion, development and strengthening of marine science in the region and a key factor for coastal and marine resources management. There is no doubt of the strategic, economic and cultural importance of the seas and coasts to a region like the Caribbean surrounded as it is by 28 states and territories, including 13 continental states and territories. The Region also has 31 small islands that deserve special attention.

The Caribbean experience has demonstrated the fundamental and key role to be played by sciences and services in making wise, timely and sustainable decisions.

The viability of marine and coastal research in the region is sustained by national capacities and by the level of cooperation. The complexity and interdisciplinary nature of modern science makes it even more important to connect their results to the policy-making process in environmental and natural resources management.

Today, it is clear that the achievement of national goals in marine issues and the capacity of participation in international programmes are interdependent. It is almost impossible that a country can be integrated to a Regional or Global programme if it has no national capacity for the solution of its own high-priority issues, in particular, for sustainable marine and coastal resources management. The experience of IOCARIBE in its many years of regional cooperation has demonstrated this fact through the gradual creation of a capacity that, although still insufficient, constitutes a good opportunity in marine and coastal resources rational use.

The last 30 years in particular, have been rich in programmes and activities developed by IOC/UNESCO, the Caribbean Environmental Programme of UNEP and other governmental and non-governmental international organizations. It is worth affirming that participating states are increasingly aware that it is only through unity, integration and cooperation, and with a true coordination of efforts and resources, that our countries will be able to address the many complex scientific and administrative issues that oceans and coastal resources demand.

A comprehensive analysis of some successful and unsuccessful regional cooperation and capacity building programmes indicates the existence of a group of basic principles:

- The programme concept must be based on the existence of a common purpose among participants, and related to an agreed theme.
- There should be a high national priority to ensure the necessary national support and follow-up.
- A minimum human and material resources capacity should be available at national level, or there should be the necessary political will to develop and to put that capacity into the national and regional interest.
- A recognized individual and/or institutional leadership should be secured in order to allow proper guidance and development of the whole programme. The leadership has to demonstrate a proven scientific and moral authority, not imposed by any reason of economic power or institutional and material development.
- The existence of a promoting, implementing and coordinating national and international institution or body should be secured and developed.
- The financial and in-kind contributions from international donor agencies and organizations, and also from participating countries, should be sustained over the long term.

Figure 15.9 Diving at all levels of experience is a major attraction in the Caribbean. © Olga Khoroshunova/Shutterstock.com.

Science plays a strategic and decisive role in the whole process. It is the basis of all phases in natural resources management and in the improvement of the quality of life. However, there still exists a lack of attention and integration in all marine programmes of the region. There are relatively more workshops, science meetings, other general activities and projects than the available human resources can use rationally. Of course, poor economies and the resulting lack of funds are highly negative factors. But it is worth pointing out that this situation is exacerbated by the lack of prioritization and the absence of the appropriate programmes, integration and coordination. Therefore, it is essential to:

1. Elaborate capacity building programmes with high standards of integration at national and regional levels that are viable in relation to the available resources in terms of funds, personnel and institutions; and that can be efficiently implemented, executed, managed and systematically evaluated.
2. Gradually achieve better communication and coordination among the different scientific and services programmes at the national, regional and global level.
3. Establish a regional coordinating body or mechanism as a proper means for coordination, rationalization and multiplication of efforts and resources of the

different governmental and non-governmental organizations involved, with clear definitions of their mission, functions and facilities.

4. Establish and improve systematic evaluations of the implementation and operation of programmes, as well as assessing their impact at national, regional and global levels.

Capacity development is a fundamental condition for scientific advancement and both highly depend on the level and conditions for effective national and international cooperation. These elements, developed on the basis described above, constitute the fundamental pillars for the intelligent and sustainable management of the marine and coastal resources.

Final comments

CICAR and IOCARIBE have marked the history of the development of marine and coastal sciences in the Caribbean region. Their programmes and activities have precipitated the creation of national infrastructures, including the development of human resources during the last 40 years. The experience gained can nourish the present and future marine and ocean policies at international and national scales.

The future role of IOCARIBE is unquestionable. It is time to recognize that only concerted action will allow us to manage better the common heritage made up of the ocean and coasts; and to identify answers and solutions to our marine environmental uncertainties and issues to benefit all states of the Caribbean region.

16 Oceans, science and governments in Africa

JUSTIN AHANHANZO AND GEOFF BRUNDRIT

Justin Ahanhanzo is coordinator of the Global Ocean Observing System in Africa (GOOS-AFRICA). He is also the Team Leader of the UNESCO programme on the Applications of Satellite Remote Sensing for Integrated Management of Ecosystems and Water Resources in Africa. He is based at the IOC/UNESCO Headquarters in Paris.

Geoff Brundrit was the Founding Chair of GOOS Africa. He has also been Chair of the GOOS Capacity Building Panel and a Core Member of the GOOS Scientific Steering Committee. At home in South Africa, he is Emeritus Professor in Physical Oceanography at the University of Cape Town and Special Adviser on Oceans and Climate Change in the Department of Environmental Affairs of the National Government.

The majority of African countries became independent some 50 years ago. This was a victorious culmination of several centuries of struggle for freedom and liberation from slavery and colonial oppression, and for self-determination. Many challenges were ahead for these newly established African Governments. There is no easy way for rebuilding a nation. New infrastructures were needed to equip the continent with unt their national culture, needs and priorities. More importantly, they had to educate new generations of citizens capable of running national structures and superstructures.

Because African countries are well endowed with one of the richest coastal oceans in the world, ocean sciences were, and remain, an important part of their national

Figure 16.1 Fishermen preparing to launch a pirogue in Benin. © Mike Markovina/Marine Photobank.

agenda. The newly independent African nations placed high hopes and confidence in international organizations to support their development efforts, for example the Intergovernmental Oceanographic Commission of UNESO.

There has been progress. Today all the African coastal states have national oceanographic research institutions. However, the level of ocean sciences in Africa reflects the overall development of scientific and technology infrastructures on the continent.

Africa is a continent surrounded by three oceans and has two continental seas. There are 44 African coastal countries, giving African nations an important position in global ocean science programmes, and indeed in global ocean geopolitics.

Historically, the control of ocean access by imperial powers led to the exploitation of Africa through slavery and colonization and the devastation of its own indigenous economy, population and socio-political foundation, over several centuries. The same command of navigation and ocean technology continues to play a role in the competition for African resources and their exploitation for the benefit of overseas economies.

Oceans and sustainable development

The ocean exerts a considerable effect on the lives of African populations and the economies of African countries. The ocean processes directly affect ships, offshore platforms, coasts and harbours, the food fisheries, marine trade and recreation, and many activities. The ocean's very large but indirect effect on weather and climate can impact lives, property, infrastructure and the economy, through droughts, floods, tropical cyclones and storm surges. Both the direct and indirect impacts of the ocean on humans are becoming more serious because of the steady migration

Figure 16.2 Plastic polymer pollution on a beach in Cape Town, South Africa, after a storm. © Maleen/Marine Photobank.

of people to the coast, and the growth of coastal mega-cities. In 2015, Lagos, Nigeria, is expected to be the third largest city in the world with a population approaching 25 million.

Marine economic activities worldwide account on average for some 5% of Gross Domestic Product. For many countries in Africa, this percentage is much higher (up to 7–10%), because of the large and growing reliance on coastal and marine living resources; offshore and coastal oil and gas and minerals; shipping and trade; and coastal tourism. The percentage is even higher (up to 50% or more) for countries like Angola, Equatorial Guinea, Gabon and Nigeria, which are particularly reliant on offshore oil and gas.

To this should be added the substantial indirect effects of the ocean on the economy, through its role in weather and climate, combined with the dependence of agriculture and energy on water supply. Many economies and societies in Africa can be considered marginal and so are particularly vulnerable to climate changes.

The potential value of the oceans for Africa may be well understood, but the benefits are still far from being fully realized. President Nelson Mandela clearly emphasized the urgent need for coupling sciences, management and human development *goals* at the 1998 Pan-African Conference on Sustainable Integrated Coastal Management (PACSICOM) in Cape Town when he stated '*Africa's long and beautiful coasts and the abundance of marine resources can contribute to improve economic, food and environmental security for the continent. These coastal and marine resources, like the rest of Africa's environmental resources, continue to be exploited in a manner that does not benefit Africa and her People. This is a paradox of a people dying from hunger, starvation and poverty when they are potentially so rich and well endowed*'.

Figure 16.3 Snorkelers in South Africa peer anxiously into the water as they prepare to enter the water and swim with sharks for the first time. Getting people to do this always takes some persuasion; after swimming with these animals snorkelers understand why they must be conserved. © Sijmon de Waal/Marine Photobank.

Ocean issues of public concern in Africa

Africans are concerned with ocean-related activities that directly affect the quality of their life and the health of marine environment and related ecosystems. In common with governments worldwide, African governments are concerned with many marine issues, such as:

- Civil protection, security and defence: coastal and offshore engineering essential to support industry and coastal protection.
- Monitoring of ocean trade and the management of ports.
- Safeguarding the security and safety of sea-goers.
- Tracking and management of coastal pollution including oil spills.

Figure 16.4 A Great White shark breaching at Seal Island, South Africa. © Dr Dirk Schmidt/ Marine Photobank.

- Impacts of floods and droughts on human life and settlement.
- Impacts of floods on infrastructure and health.
- Impacts of drought on agriculture.
- Forecasting of droughts and floods.
- Enhancing preparedness for and dealing with natural disasters, such as tropical cyclones and storm surges.
- Enhancing the preparedness of health carers for climate-induced disease outbreaks.
- Sustainable use of living marine resources in the coastal zone and coastal seas.
- Management of key habitats and ecosystems in those areas.

Although these are issues of global concern, many have aspects that are of special significance to Africa. Some of these are discussed here.

While the general public are well aware of the direct impact of the oceans on the coastal economies, much less is known about the ocean's very large but indirect effect on weather and climate. Rainfall in much of southern Africa is influenced strongly by the El Niño events originating in the equatorial Pacific Ocean. During El Niño years, droughts prevail in much of southern Africa, causing widespread economic hardship.

Fluctuations in the Tropical Atlantic Dipole in the equatorial Atlantic bring warm surface waters north as far as Mauritania, feeding the rains that water the Sahel every decade or so. Fluctuations in the Indian Ocean Dipole bring warm surface waters to the coasts of Kenya and Somalia, feeding the rains that drench east Africa also every decade or so. Dry conditions are typical in the intervening period when the surface waters in those areas cool.

The southern end of Africa experiences wet conditions when the area is under the influence of the warm waters carried east in the Circum-Antarctic Current by the Circum-Antarctic Wave approximately every 9 years; between these times cool waters bring drier conditions. The northern margins of Africa lie under the influence of the North Atlantic Oscillation, which brings wet conditions during its negative phase and dry conditions in its positive phase.

Oceans support not only marine life but also life on land, through rainfall. Rainfall on land supports hydroelectric power and the growth of crops; however, excess rainfall can damage infrastructure, such as roads and bridges, making the distribution of food difficult. Excess rain and drought also affect human health, not least through effects on disease vectors, such as malaria.

Cyclones and lesser storms are the chief agents of coastal erosion, a major threat to infrastructures along the African coastline. The tropical regions of the Indian Ocean generate tropical cyclones that can create havoc when they make landfall on the east African coast. The cyclones derive their energy from the ocean. As they strengthen when passing over patches of warm water, or weaken over cool water, ocean information is essential in forecasting their strength and the likelihood of damage. When storms at sea or cyclones make landfall, their potential for damage is greatly heightened by coincidental high tides, which create massive storm surges to flood low-lying coastal areas. Tropical cyclones kill people and cause damage amounting to tens or hundreds of millions of dollars. On average there are about 10 cyclones per year during the cyclone season between November and May.

Large-scale ocean currents are certainly important for African economies. The Canary Current in the north-west, the Somali Current in the north-east, the Guinea Current in the West and the Benguela Current in the south-west are all associated with upwelling processes that stimulate high marine productivity and support considerable fisheries. The Agulhas Current is a southern equivalent of the Gulf Stream, transporting massive amounts of heat away from the equator and, indeed, feeding some into the Atlantic. Hence it can ultimately affect the climate of north-west Europe, emphasizing the global nature of ocean influences.

Figure 16.5 Africa is planning for an ocean future. © Jeff Gynane/Shutterstock.com.

African responses to ocean issues

African governments have responded to this general and growing awareness of the socio-economic importance of African seas. Countries and regional institutions have recognized the importance of establishing monitoring programmes to facilitate the generation of oceanographic, meteorological and hydrological data. A number of projects have been funded in African coastal waters under the Abidjan and Nairobi Conventions established in the 1980s. Joint initiatives among African nations have been undertaken since 1992 in the field of meteorology and its applications in development through the setting-up of the African Center for Meteorological Applications for Development (ACMAD). These initiatives are directly funded and supported by African governments.

Most recently, the development of the Large Marine Ecosystem (LME) projects, funded by the Global Environment Facility (GEF) of the World Bank, focus on the development of management tools for coastal and international water resources in the Gulf of Guinea, the Benguela and the Aghulas–Somali Current regions. The focus of these projects is naturally ocean-directed. Taking into consideration the land–ocean–atmosphere interaction, a synergy was established with land-oriented projects including the PUMA, a joint EUMETSAT and WMO project aimed at equipping meteorological offices with the ability to use data from the METEOSAT Second Generation Satellite. However, there is a lack of a credible Africa-wide data and information management system to provide decision-making support to all the stakeholders in the region, although the Flanders sponsored ODINAFRICA project provides some assistance for the establishment of national oceanographic data structures in a number of African coastal countries. Indeed, the existing projects do not as yet constitute a sustained and integrated operational system. There is a gap to be filled.

Very aware of the importance of marine and coastal environments in Africa, the African governments developed national and regional frameworks for the management of their coastal ocean space and resources. The regional frameworks include the Abidjan and Nairobi Conventions for Cooperation in the Protection and Development of the Marine and Coastal Environment, of the West and Central and East African regions respectively.

Within the framework of the African Union development plan, known as the New Partnership for Africa's Development (NEPAD), sectoral priority areas are included such as: infrastructure, especially in computing, communications and energy; human resources, capacity development initiative, including education, skills development and reversing the brain drain; Agriculture; the Environmental Initiative; Culture; and Science and Technology platforms. However, more precisely, it is the Environmental Initiative and Science and Technology platforms that are of greatest interest to the promotion of ocean sciences, technology and infrastructures. The Heads of African States put a special emphasis on the Environmental Initiative including marine and coastal ecosystems management, and pollution, global warming and climate change; Wetland Conservation; Combating Desertification; Invasive Alien Species; Cross-border Conservation Areas; and Environmental Governance and financing, thus demonstrating their commitment to the sustainable development of the African coastal seas and oceans and the preservation of this unique environment.

It is important to underline that Africa is the only region where a continental Summit of Heads of States adopted a regional Environmental Action Plan. GOOS-AFRICA, as a key contributor to the NEPAD marine and coastal programme, was subsequently endorsed by the Heads of States. Within the framework of NEPAD, the project on Shoreline Change and Adaptation to Climate Change in West Africa developed and received substantial implementation funding from the Global Environmental Facility of The World Bank. The expression of political will was an important first step that raised the profile of environmental issues including coastal and marine aspects. Further optimism followed two environmental Donor Conferences, December 2003 in Algiers, and March 2005 in Dakar. Nevertheless, progress with funding and implementing projects, including some 30 marine and

Figure 16.6 Fishing boats on the beach in Senegal. © Kirsz Marcin/Shutterstock.com.

coastal related projects within GOOS-AFRICA, has been slow. The low rate of project implementation is due to insufficient financial commitment from African Governments, as they appear to be relying on overseas donors to fully fund their domestic projects. There is now an urgent need for political commitment of African Governments to establish African funding mechanisms which will fast-track the implementation.

African Governments are parties to the relevant United Nations Conventions including the following:

- The United Nations Convention on the Law of the Sea (UNCLOS)
- The United Nations Convention on Biological Diversity (UNCBD)
- The United Nations Framework Convention on Climate Change (UNFCCC)

Again, a regional cooperation through a common African voice at the level of the Heads of States was decisive to ensure a timely submission of claims on the Limits of continental shelf, which lie beyond 200 nautical miles from the baselines from which the breadth of the territorial sea of the two States is measured. In fact, the February 2008 Assembly of the African Union in Ethiopia adopted the Decision on the Extension of the African Continental Shelf and Climate Change that called on African Coastal Member States to speed up the process of preparing and submitting the claims for the extension of the limits of their continental shelf, requested the AU Commission to assist Member States in this regard, with the view to meeting the deadline of 12 May 2009, and also requested the African Union Commission, in collaboration with Member States, to engage in a process to designate an eminent African personality as Special Envoy of the Union in charge of maritime and climate change matters, and finally called on the specialized agencies of the United Nations system including IOC/UNESCO and UNEP to provide the coastal

Member States with the assistance required to conduct the necessary studies for submitting applications for the extension of their continental shelf. As a result, it is worth noting that on 1 December 2008, the Republic of Mauritius and the Republic of Seychelles were the first African Islands to submit to the Commission on the Limits of the Continental Shelf, in accordance with Article 76, paragraph 8, of the United Nations Convention on the Law of the Sea, a joint claim containing information on the limits of the continental shelf appurtenant to the Republic of Mauritius and the Republic of Seychelles, respectively, which lie beyond 200 nautical miles from the baselines from which the breadth of the territorial sea of the two States is measured in the region of the Mascarene Plateau. Other coastal countries are following the process.

Operational oceanography in Africa

In the near future operational oceanography worldwide will be called upon to provide similar services to those presently available for meteorology. African countries are increasingly active in operational programmes to address the needs of global conventions. African nations embarked in the development of the operational oceanography through the implementation of the Global Ocean Observing Systems in Africa (GOOS-AFRICA). They established Pan-African technical and scientific teams supported by international networks of institutions specialized in a range of various ocean-related sciences including: *in situ* ocean observations of sea level; data collection and data centres; coastal satellite applications and remote sensing; modelling and data assimilation; forecasting products and services; and end-user interaction and public awareness.

This integrated system will provide relevant information and services including the following:

(1) Accurate descriptions of the present state of the sea and its contents including living resources and contaminants.
(2) Continuous forecasts of the future conditions of the sea and its contents for as far ahead as possible.
(3) Long-term data sets and information showing trends and changes, including the effect of the ocean on climate and climate change in coastal regions and vice versa.

The observing systems around Africa comprise:

(1) Voluntary observing ships collecting oceanographic and meteorological information.
(2) Ships of opportunity collecting subsurface temperature and salinity data.
(3) Surface drifting buoys collecting ocean surface and meteorological data.

Figure 16.7 Larger fishing vessels in a harbour near Cape Town, South Africa. © MaxPhoto/Shutterstock.com.

(4) Fixed buoys in the equatorial Atlantic (the PIRATA network) collecting surface and subsurface ocean data and meteorological data.
(5) Coastal and island tide gauges operations.
(6) Satellite measurements for a wide range of ocean properties.
(7) Profiling Argo floats collecting subsurface temperature and salinity data.
(8) National research vessels collecting oceanographic and fisheries data.

These observing systems are substantially driven from outside the continent and appear to have very little interaction with local laboratories and stakeholders. Regrettably, therefore, they are not intrinsically sustainable. The dilemma is that only a few African countries have developed a national ocean policy and research strategy. Long-term African national vision is needed to build and maintain sustained ocean observing systems.

To a large degree, the previously funded coastal margin projects were developed in the era preceding the very substantial advances in numerical modelling and forecasting that have been stimulated by the massive increases in computer power of recent years. Thus it has only now become possible to run ocean models at a high enough degree of resolution to provide really useful information to coastal managers. In addition, previous projects have not sought to capitalize on the integration of the massive amount of presently available ocean and atmospheric data collected by remote sensing from satellites with the information available from *in situ* ocean instruments, and to assimilate these data into the new generation of forecasting models. Subsequently, African countries are now in the process of building their own national and regional scientific and technology infrastructures and capabilities that enable them to tackle the ocean modelling activity in a way never before attempted or achievable.

There are significant efforts at national and regional level in Africa to establish world class modern scientific and technology infrastructures including high-performance computing facilities, the ultimate prerequisite for the development of operational oceanography. A few examples of these African initiatives follow.

Southern Africa

- ***The Centre for High-Performance Computing (CHPC)***. This is the initiative of the Department of Science and Technology of the Republic of South Africa. This world class numerical facility defines itself as a Pan-African platform for advanced research and education in earth system sciences including the oceans and climate change studies. During a recent visit of the GOOS-AFRICA Coordinator to the Centre in July 2009, the Centre Management offered to make the facility available for the promotion and development of the operational oceanography in Africa through the GOOS-AFRICA Framework including joint research programmes and training. The CHPC organized annual workshops on the advances in high-performance computing.
- ***The African Centre for Climate and Earth Systems Science (ACCESS)***. This is the initiative of the Department of Science and Technology of the Republic of South Africa aimed at supporting advanced research and training in climate and earth system sciences including the oceans–land–atmosphere interactions. Since 2008, ACCESS has organized annual workshops for African master students from various scientific and academic backgrounds on 'How to build a habitable planet' and since 2009, has instituted an annual high-level research conference on advances in operational oceanography and climate studies. Both events are hosted at the University of Cape Town, the place where three oceans meet.
- ***SADC Centre for Meteorology and Agriculture based in Botswana***. This is the initiative of the East African Economic Commission.
- ***SADC Fisheries Center***. Based in Namibia, this is the initiative of the Southern Africa Development Community.

Eastern Africa

- ***IGAD Climate Prediction and Applications Centre***. Based in Kenya, this is the initiative of the East African Economic Commission.
- ***Regional Centre for Mapping of Resources for Development (RCMRD)***. This is the initiative of the Economic Commission of the East African Development Community.

Figure 16.8 Fishermen set out to sea in The Gambia. © Trevor Kittelty/Shutterstock.com.

Western Africa

- ***The African Centre for Meteorological Applications for Development (ACMAD).*** This is the initiative of the African Governments supported by the Economic Commission for Africa. The Centre, based in Niger, provides meteorological information and weather forecasts for the region. It also serves as a training centre for national meteorologists.
- ***Regional Center for Climate Modelling and High Performance Computing.*** This is the initiative of the Government of Benin to provide advanced scientific and research infrastructures in the West Africa region for climate and ocean modelling.
- ***Regional Centre for Training in Aerospace Surveys (RECTAS).*** Based in Nigeria, this is the regional centre under the auspices of the Economic Commission for Africa.

The above initiatives contribute to the development of satellite remote sensing applications and ocean and climate modelling and forecasting in Africa. Improved regional ocean forecasts products and services will be provided in a timely manner to a wide range of users and stakeholders.

Building ocean sciences in Africa

All African coastal countries have established national ocean research institutions, and these are at various stages of development. In addition to that, several African universities developed research and education programmes in earth, ocean and atmosphere sciences and climate studies. The UNESCO, through the IOC, encourages the establishment of UNESCO Chairs in marine sciences and oceanography in a number of African countries. The UNESCO Chairs provide a visibility to national efforts and promote South/South and North/South Cooperation through joint research and education programmes. However, existing efforts and programmes are fragmented, uncoordinated and lack long-term national and regional vision and perspective.

There are very few UNESCO Chairs in Africa and even fewer in marine-related fields. This limited number includes:

Algeria: Chair in Prospective and Integrated Coastal Zone Management established in 2005 at the Institute of Marine Sciences and Shoreline Management (ISMAL).

Benin: Chair in Mathematical Physics and Applications established in 2006 at the University of Abomey-Calavi. Recently, MSc and PhD diplomas were established in geographical information system (GIS) data manipulation, and Applications of Satellite Remote Sensing for Integrated Management of Ecosystems and Water Resources and in Physical oceanography for the entire West African region.

Morocco: Chair in training and research in marine sciences established in 1994 at the University of Chouab Doukkali, El Jadida and chair in natural gas and environmental management and sustainable development established in 1997 at the University Mohamed V in Rabat.

Mozambique: UNESCO Chair in Marine Sciences and Oceanographic Issues established in 1998, Universidade Eduardo Mondlane in Maputo.

It is worth noting that the Commission of Sciences, Technology and Human Resources of the African Union is in the process of establishing regional Pan-African Universities with focal areas including earth observations and space sciences and technology, climate and oceans systems.

Support from the International Assistance for Capacity Building in Ocean Sciences in Africa has until now focused on short training workshops and use of consultants, mainly from the donors and overseas countries, with less support given to local infrastructure and institutional building. The unfortunate consequence is that the available expertise and capacity in Africa are not fully used and promoted. Recent policy development in Africa strongly recommends that international and overseas initiatives MUST recognize existing African

Figure 16.9 Relaxing on an unpolluted beach in Tanzania. © BlueOrange Studio/Shutterstock.com.

capacity including scientific and technological infrastructures, programmes and expertise, and should utilize, and ultimately improve, the existing framework and personnel. There is an urgent need for building infrastructure and institutions with particular emphasis on the development of early warning systems, for mapping resources, for understanding trends in environmental change and for predictions of extreme events. These services are essential in decision-making for coastal managers concerned with basic needs. The key for a sustained capacity is the inclusion of ocean sciences in the regular university training and education system with a multidisciplinary approach including *in situ* and remote data collection, processing, analysis and interpretation, together with the development of modelling and forecasting skills leading to the sustainable provision of ocean products and services. There should be an expansion of the use of new telecommunications technology for the delivery of services.

Capacity development should therefore take on an operational profile, making use of robust observing and forecasting systems, appropriate to Africa.

The future of ocean sciences in Africa depends on the ability of African governments to build the future generations of African champions in ocean sciences based on the following facilities and mechanisms:

- On-the-job training for individuals linked to pre-operational pilot projects.
- Empowerment of regional capacity in earth observations with particular emphasis on oceanography and earth system sciences studies with a strategy combining the provision of training, equipment and operational support.
- Regional training programmes (MSc/PhD) with particular emphasis on operational oceanography, modelling, climate change and earth system sciences.

The success will mainly depend on the volume of national and regional investment for attractive career development in an enabling conducive environment for scientific research and technology development. African stewardship can then become a key contributor to the global knowledge in ocean sciences.

Finally, the three pillars in international cooperation for a future capacity building in the so-called developing world in general, and in Africa in particular, should be the following: (1) a grassroots approach; (2) ownership and commitment; and (3) leadership from the local, national and regional stakeholders.

The power of cooperation

Regional and international cooperation is fundamental to ocean governance and management because the ocean processes and events are commonly a function of earth, oceanic and atmospheric processes on regional and global scales. African governments are working to deliver participation on all scales. Participation is part of a global trend, with the establishment of GOOS and GCOS supported by the UNESCO and its IOC, UNEP, GEF and UNDP and the formation, by WMO and the IOC, of the Joint Technical Commission for Oceanography and Marine Meteorology (JCOMM).

Sea life, pollutant dispersion and storm damage are transboundary issues affecting regional groups of countries, and requiring close collaboration and cooperation at regional and international levels and between oceanographic, meteorological and civil engineering agencies. The oceans become polluted from time to time by land runoff and by oil discharged accidentally or deliberately at sea or from wrecks. Pollution arising from one source often becomes the concern of many countries and communities as it is spread far and wide by the sea.

Under such circumstances it would be impossible for any single organization or country to establish the necessary infrastructures and make available the required human resources to supply the relevant data and services needed to deal with these issues in an efficient and cost-effective manner. The most appropriate response is regional.

The African nations with their Pan-African coordinating mechanisms in charge of promoting operational oceanography in Africa fully appreciate the power of cooperation and are well aware that the scale of investments and complexity involved in the building up of operational oceanography systems would overwhelm the capacity of any single African nation. Moreover, the transboundary nature of the ocean phenomena requires international jurisdiction and a legal framework. Consequently, regional African Cooperation and international partnerships have been established to foster the implementation of operational oceanography.

Figure 16.10 A penguin surveys the sea in Cape Town, South Africa. © Shuttervision/Shutterstock.com.

Conclusion

The challenges facing African countries, if they are to fully harness the societal benefits of their oceans resources for poverty alleviation and eradication in the continent, lie in a series of questions including the following:

1. What are the needs of African countries in a 20- and 50-year period regarding ocean resources including fisheries, oil and gas?
2. What is the knowledge needed in Africa regarding ocean sciences and technology in order to respond to the above question?
3. What are the institutional infrastructure and human capital needed to respond to questions 1 and 2 above?

African nations should ultimately empower their national scientific and technology infrastructures in order to reinforce their ability to generate useful local, national and regional ocean information products and services for national development. Furthermore they should strive to contribute to the objectives of national ocean legislations, regional and global Conventions and Agreements such as the Rio Conventions, notably, the UNFCCC and the CBD.

African governments should aim to make African nations self-sufficient in using ocean observing systems to advance local economies. Therefore, their vision for Africa **must be** to establish a completely integrated and fully operational African ocean observing and forecasting system through:

- the development of institutional and human capacity in Africa;
- the development of partnerships, both existing and new, and to work **together** in achieving this vision and, in this way, help to keep alive the dream of African youth that hungers for knowledge and for a better life in Africa.

In the future of operational oceanography in Africa, GOOS-AFRICA intends to facilitate and coordinate united action towards the achievement of the above vision, and using the integrated and operational ocean observing system to contribute to the future well-being of the African people. GOOS-AFRICA will help to ensure an effective response to natural disasters that come from the ocean, and to the challenges of climate change on the coast and in the oceans surrounding Africa. This will be achieved through creating awareness and effective policies built on the increased knowledge from the ocean observation system. These actions will form an important African contribution to the Implementation of the GEO/GEOSS and the IOC/UNESCO Plan for Integrated Coastal Observations of the Global Ocean Observing System.

As African nations empower their national and regional scientific, technology and industrial infrastructures, a future Africa can proudly contribute its own voice on global warming and climate change issues.

Part VI Intergovernmental Agencies and Science: Preface

In the previous section the authors showed the wide range of applications of marine science, and the different emphasis given to different aspects, coastal and open ocean, in each region. This diversity is further reflected in the range of intergovernmental agencies concerned with the sea and its management, both within the UN system and in other governmental and non-governmental groups.

When dealing with an international commons as large and important as the ocean, this diversity is expected. Many of these deal directly with specific areas such as the Food and Agriculture Organization for fisheries and the International Maritime Organization for transportation; others are more broadly concerned with ocean programmes, such as the involvement of the World Meteorological Organization in weather and climate forecasting, and the United Nations Environment Programme for the environment. A more extensive list includes the Seabed Authority, the World Health Organization, the International Atomic Energy Authority and of course the UN in New York for the Law of the Sea and its application. The following section, therefore, is only a sample of the intergovernmental governance activities of the UN related to oceans. Of all these however, the Intergovernmental Oceanographic Organization of UNESCO is the only UN organization charged solely with the delivery of ocean science and services.

Non-UN organizations also flourish and make substantial contributions to the effective management of the oceans, based on good science. These include the International Council for the Exploration of the Seas based in Denmark, founded over 100 years ago, and the International Hydrographic Organization, based in Monaco since 1921. Diversity is often a sign of collective strength, and for the oceans, the wide range of interested bodies and agencies adds up to a considerable body of human marine intervention and activity. However, this diversity also brings problems. This breadth of ocean interests is often a barrier to a coordinated approach to ocean governance. The sum of the activities is often less than the component parts.

Competing and overlapping responsibilities among the international agencies is also reflected at the national level. Within individual nations, the ocean mandate is inevitably spread over ministerial portfolios representing transportation, health, energy, defence, environment, and even culture and recreation. Most often, conflicts of marine interests are not resolved at a national level with the result that differences can be reflected internationally through delegates to the respective intergovernmental meetings reflecting their own specific positions instead of a coordinated national approach. Bilateral and multi-agency cooperation on issues of mutual or common interests can therefore be hampered by this lack of communication between and amongst delegates from different national departments, but representing the same government.

Governments recognize and acknowledge these concerns. Though acknowledgement of a problem is not a solution, it is a necessary first step! There are useful current initiatives within the UN system as a whole, encapsulated in the 2006 'Delivering as one' manifesto. For the future of the oceans, the recent (2003) creation by the UN of the interagency UN-Oceans is a work in progress. It is composed of the relevant programmes, entities and specialized agencies of the UN system, with the goal to establish an effective, transparent and regular interagency coordination mechanism on ocean and coastal issues within the UN system. Another significant recent development is the 'Open-ended Informal Consultative Process on Oceans and the Law of the Sea' (ICP) held in New York each year with results feeding into the UN General Assembly. These are indications of the attempt by governments to address the need for better liaison and coordination. In a sense the ICP can serve as a coordinated customer for ocean management and especially for ocean science.

The following articles have been invited from the major UN agencies that have scientific ocean needs and interests, and from the IHO; all have worked with the IOC over the past decades. Although each has its own set of priorities, there are common interests, and a concern for the development of coordinated high-quality marine knowledge and information which has encouraged and fostered mutual cooperation.

17 The Food and Agriculture Organization

RAY C. GRIFFITHS

Ray C. Griffiths took an MA at Oxford University. His employment record comprises: University of Toronto; Fisheries Research Board of Canada; Scripps Institution of Oceanography; government of Venezuela; and FAO, first as a field expert, then seconded to the IOC in 1973 and, in 1988, as Secretary of the General Fisheries Commission for the Mediterranean.

The Food and Agriculture Organization of the United Nations, with its Headquarters in Rome, provides a neutral forum where all of its 191 Member States meet as equals to debate policy and negotiate agreements on relevant intergovernmental cooperation. The FAO helps developing countries and countries in transition to modernize and improve agriculture, forestry and fishery practices and thus to improve human nutrition.

By virtue of its wide mandate the FAO is concerned with all the principal natural sciences insofar as they relate to the exploitation of natural living resources and, in the present context, fishery resources. For more than 100 years, and particularly since the early 1950s, fishery science has played a central role in the management of capture fisheries. Many disciplines have been involved: biology and ecology, as resources were discovered; and technology, as fisheries expanded and modernized and as excessive fishing capacity and overfishing led to subsequent, often spectacular, fishery collapses. This experience laid the basis for resource management, but also for the application of social science and, more recently, political science to what may be called fishery governance, which is now backed scientifically by FAO's ecosystem approach to fisheries.

The UN Convention on the Law of the Sea calls for States, in determining the allowable catch and establishing other conservation measures for the living resources, to take measures based on the best scientific evidence available (Article 119).

Consequently, research is the cornerstone of effective fishery management and an important element of the future planning of capture fisheries and of aquaculture.

Improvements in fishery science are still needed and depend mainly on:

- a better understanding of ecosystem structure, functioning and resilience;
- the development, standardization and wide application of indicators of fish stock and fishery sustainability; and
- an increased application of social sciences.

Fishery managers are nevertheless often required to make decisions in the face of a high degree of uncertainty due mainly to:

- the statistical variability associated with the sampling (catching) of largely invisible and moving organisms, often over vast areas of ocean;
- the natural variability and inherent complexity of the ecosystems of which fish stocks are a part as well as the rapidly changing preferences, expectations and motivations of human populations, and the incomplete understanding of the functioning of the fishery system and hence of its mathematical representations; and
- the inability to predict the reaction of the participants to fisheries management measures.

The perspective of climate change has added an additional and complex dimension to all these uncertainties, calling for an even more precautionary approach to fisheries and a type of science more able to face uncertainty.

To be effective, fishery management is significantly dependent on the extent to which it is informed by a clear understanding of the factors that determine the abundance, dynamics and resilience of fishery resources, and on an understanding of the structure and socio-economic dynamics of the fishery sector. The achievement of such understanding requires the collection and analysis of data and information. Since experimentation is not practicable, the conception and testing of complex models is essential to:

- elaborating fishery development and management (including stock rebuilding) plans;
- establishing indicators of sustainability and understanding their behaviour;
- developing cheaper and more environmentally friendly technology;
- testing precautionary approaches to the protection of fishery resources and fishing communities;
- improving food processing and other forms of value adding; and
- testing the impacts of fishery regulations.

Many important fisheries involve more than one country and are regulated by regional fishery commissions or other legal arrangements. In some instances, a

Figure 17.1 A small fishing vessel in the Baltic Sea. © LianeM/Shutterstock.com.

commission (e.g. the Inter-American Tropical Tuna Commission) has its own means of generating the scientific advice that it is mandated to provide to its Member States. In others (e.g. the General Fisheries Commission for the Mediterranean), the scientific information is assembled at national level and submitted by the Member States to the Secretariat who will use it in assessment and advisory working groups. In either case, technical staff (either independent or national) assemble and analyze the data, evaluate the likely outcome of alternative decisions and make recommendations accordingly; but the final decisions are made by States (or the European Union) and their fisheries authorities. Historically, short-term national socio-economic interests have usually prevailed over long-term conservation requirements, particularly where scientific understanding has not yet been conclusively established (as is often the case in the developing world) but also when high-quality advice was available (as in the North Atlantic).

Although the mandates of the UN specialized agencies are essentially operational, given the relatively limited scientific expertise available to them within their own secretariats, some of them have created their own advisory bodies. The FAO has established a large number of Regional Fishery Bodies since 1949, most of them before the UN Convention on the Law of the Sea was adopted in 1982, with research and advisory mandates, over EEZs and the high seas. Most of them remained in function after the UNCLOS adoption by will of their Member States. At the global level, the FAO created its Advisory Committee on Marine Resources Research (ACMRR) in 1962. This Committee concerned itself with most of the major fishery issues, including fish stock assessment, research and management, fishery economics, and aquaculture, but it was particularly concerned with the fisheries for marine mammals until the establishment of the International Whaling Commission. Since then, the FAO deals with marine matters only when there is direct interaction with fisheries.

Around this time there was a growing concern that oceans needed to be studied in a coordinated way to address many issues of concern to UN Agencies. In 1966, the General Assembly of the UN requested the Secretary-General to make proposals for an expanded programme of international cooperation in the study of the oceans; this was done in collaboration with UNESCO/IOC and the FAO. The UN General Assembly endorsed the Secretary-General's report and welcomed a proposal for an International Decade of Ocean Exploration (IDOE), as the initial acceleration phase of the expanded programme. Pursuant to the UN Resolutions, a Joint Working Party on the Scientific Aspects of International Ocean Research, in 1969, produced a report titled 'Global Ocean Research', which identified four major fields of international endeavour for immediate consideration by Member States:

- Ocean Circulation and Ocean–Atmosphere Interactions
- Life in the Oceans
- Marine Pollution
- Dynamics of the Ocean Floor

With this report as a basis, the UN General Assembly endorsed a Long-term and Expanded Programme of Oceanic Exploration and Research (LEPOR), in 1969. The Global Ocean Research report itself may therefore be confidently identified as the act of conception of many of the major international scientific programmes carried out in the context of the LEPOR and IDOE (1971–1980), and beyond.

The LEPOR/IDOE projects of direct interest to FAO included:

- investigation of the phenomenon known as El Niño;
- study of the physical processes connected with the upwelling in the northern part of the eastern central Atlantic;
- study of the physical oceanography of the Kuroshio and adjacent regions;
- study of the physical oceanography of the Mediterranean;
- baseline studies on the inputs, pathways and levels of synthetic halogenated hydrocarbons and petroleum hydrocarbons in selected key organisms;
- global investigation of pollution in the marine environment (GIPME);
- study of North Sea pollution;
- coastal upwelling ecosystem analysis (CUEA);

Figure 17.2 Fish for sale in the Maldives. © Alex Barron/Marine Photobank.

- assessment of living resources in the North Atlantic;
- fish stock assessment in the South Atlantic; and
- Indian Ocean fishery survey and development programmes.

In 1968, in response to the development of the role of the IOC as a joint mechanism of the UN specialized agencies that had agreed specifically to support the IOC, the Director-Generals of UNESCO and FAO signed an *aide-mémoire* specifying the nature of cooperation between the two organizations and, in particular, formalizing the secondment of a high-level FAO professional staff member to UNESCO as the liaison officer between the FAO and UNESCO. That staff member is the author of the present review.

In 1969, UNESCO, FAO, WMO, IMO and the UN created a mechanism to harmonize better their activities in the field of marine science: the Inter-Secretariat Committee on Scientific Programmes Relating to Oceanography (ICSPRO). This arrangement included the secondment of staff members from other Agencies to work within the Secretariat of the IOC, in Paris.

More specifically, and in view of rapidly growing concerns about marine pollution in the 1960s, the FAO, IMO and UNESCO decided to create, in 1968, a Joint Working Party to advise their respective Member States on all the scientific problems of marine pollution, notably with respect to biological and physicochemical effects, analytical methods and standards, and safeguards, protection and control. The FAO's main concern here was the possible effects on living marine resources and seafood contamination; IMO's was the maritime transport of dangerous chemicals, especially oil; and UNESCO/IOC's was the intergovernmental coordination of the relevant international scientific research. The independence of such a Joint Working Party from political, commercial or even specific agency interests was stressed, however. The WMO, whose main concern was the airborne transport of pollutants to the ocean, quickly became a co-sponsor, followed in 1971 by the World Health Organization, for questions of human health, and by the IAEA for the radio-ecological problems due to nuclear fuel transport and waste discharge. The UN joined in 1971. The Joint Working Party evolved quickly into the Group of Experts on the Scientific Aspects of Marine Pollution (GESAMP), which was later renamed Group of Experts on the Scientific Aspects of Marine Environment Protection, following UNEP's co-sponsorship in 1977. GESAMP is an advisory body to several agencies, and now is actively participating in the first phase (Assessment of Assessments) of the UN's Regular Process for the Global Reporting and Assessment of the State of the Marine Environment.

The FAO played a leading role in the development of activities in the marine pollution field, notably by organizing a Technical Conference on Marine Pollution and its Effects on Living Resources and Fishing (FAO, Rome, 9–18 December 1970) and the publication of the contributions thereto: Marine Pollution and Sea Life.

Regarding specific areas of cooperation between FAO and UNESCO/IOC, one of the major IOC programmes of interest to the FAO was the Global Investigation of Pollution in the Marine Environment (GIPME). The possible content of such a vast undertaking was developed at a meeting of the ACMRR–SCOR–ACOMR–GESAMP Joint Working Party on GIPME, in 1971. And, in 1972, the IOC created an intergovernmental Working Committee for GIPME. Under its auspices, with the close collaboration of the FAO and, later, the UNEP, the Working Committee pursued a wide range of inquiries, notably: river inputs to ocean systems, pollution of the ocean originating on land, atmospheric transport of pollutants, dynamics of ecosystems in relation to pollution and health of the oceans.

In more recent years, governments have become increasingly aware of two major marine ecological problems:

- The increasing frequency of harmful algal blooms (HABs), also known as 'red tides', mainly in coastal seas, possibly, though not certainly nor uniquely, due to the increasing pollution of the marine environment.
- The growing degradation of coral reef ecosystems, possibly due at least partly to global warming and marine pollution.

Under the banner of the IOC programme on International Oceanographic Data and Information Exchange (IODE), the FAO and IOC embarked upon the creation of an Aquatic Sciences and Fisheries Information System (ASFIS) in the early 1970s, and, as a derivation of it, with the collaboration of a commercial publisher (Cambridge Scientific Abstracts), the Aquatic Sciences and Fisheries Abstracts (ASFA).

While the FAO has collaborated with many scientific bodies, and specifically with UNESCO and the IOC over the years, it has progressively reduced its involvement, largely because of the growing gap between mainly ecological research and the

Figure 17.3 Sustainable fishing needs scientific understanding and political will. © J Simpson-MSC/Marine Photobank.

socio-economic reality of fisheries and, in general, of the integrated use of aquatic resources. With reduction of resources, the FAO has had to refocus its collaborations. Although GLOBEC has steered back to fishery issues, connections had been weakened. The main problems of governments, as far as fisheries were concerned, have become more socio-economic and more directly oriented towards fishery management. Significantly, the FAO's interest in the evolving Global Ocean Observing System (GOOS) is as a source of data products relevant to fishery development and management. The FAO proposed to contribute to fisheries resources monitoring and has developed the Fisheries Resources Monitoring System (FIRMS) in part as a possible component of GOOS.

Economic, social and political forces are inevitably involved as countries, in the case of shared resources, or communities within them, in the case of nationally owned resources, compete to maximize their respective shares of these resources. The starting point remains the same, however: an objective assessment of the production that can be expected from a resource and of the conditions under which such production can take place and be sustained. The future will see the development of a form of research more aware of the consequences of the complexity of the social-ecological systems of use of natural resources, their emergent properties, their sensitivity to external drivers, and the need to develop models combining soft and hard information, scientific and traditional knowledge, in highly participatory research, assessment and advisory processes and institutions.

18 The International Hydrographic Organization science partnerships

HUGO GORZIGLIA

Captain Hugo Gorziglia graduated as a naval officer in 1970, and served in the Chilean Navy as a Hydrographic Engineer in the Hydrographic and Oceanographic Service (SHOA), becoming Head of Department, Technical Advisor, Deputy Director and, finally, retired as Director in 1997. He was elected Director of the International Hydrographic Organization in 2002 and re-elected in 2007.

The Member States of the International Hydrographic Bureau (IHB), established in 1921 and now known as the International Hydrographic Organization (IHO), have pursued the aim of making navigation easier and safer throughout the world. Consequently there has always been a close synergy between the IHO and the Ocean Sciences. Indeed one of the organization's objectives articulated in the 1969 IHO Convention is: '*The development of the sciences in the field of hydrography and the techniques employed in descriptive oceanography*'.

The definition of 'Hydrography' as published in the *Hydrographic Dictionary* (IHO Publication S-32) is: '*The branch of applied sciences which deals with the measurement and description of the physical features of oceans, seas, coastal areas, lakes and rivers, as well as with the prediction of their change over time, for the primary purpose of safety of navigation and in support of all other marine activities, including economic development, security and defence, scientific research, and environmental protection*'. Therefore the IHO's contribution to ocean sciences is as wide as its imagination allows. Hydrography provides the fundamental backdrop for almost every operation or activity that happens in, on or under the sea.

The IHO, being aware of the growing need for close cooperation and collaboration in ocean sciences and recalling that the Intergovernmental Oceanographic Commission (IOC) constitutes an important coordination centre of worldwide

Figure 18.1 The Monaco Headquarters of the International Hydrographic Organisation. © IHO.

marine scientific research, decided to establish and sign a Memorandum of Understanding with the IOC on 25 January 1984, which today (updated in 2000) constitutes the reference frame for the cooperation and supportive relationship which exists between these two international and intergovernmental organizations.

One of the important decisions taken by both organizations was to cooperate in enhancing the General Bathymetric Chart of the Oceans (GEBCO) and the International Bathymetric Charts (IBC), promoting the free exchange of processed bathymetric data. The GEBCO Project started in April 1903 following His Serene Highness Prince Albert I of Monaco's offer to organize and finance the production of a new series of charts to be designated 'The General Bathymetric Chart of the Oceans'. The first edition of the GEBCO was published in May 1905 based largely on data published by the French and British Hydrographic Offices. Subsequently, with the contribution of the Oceanographic Museum of Monaco, followed in 1929 by the then International Hydrographic Bureau and later in 1970 by the IOC, five editions of the paper series of GEBCO charts have been published. These then gave rise in 2003 to a digital version called the GEBCO Digital Atlas (GDA). This modern digital product is a unique reference source for the bathymetry of the world's oceans. It contains a global one arc-minute bathymetric grid – the GEBCO One Minute Grid – and global sets of digital bathymetric contours and coastlines. This achievement, directed by the GEBCO Guiding Committee, together with the significant contribution from the two subsidiary bodies, the Sub-Committee on Undersea Features Names (SCUFN) and the Technical Sub-Committee on Ocean Mapping (TSCOM), marks out a clearly successful joint IOC/IHO project, the products of which constitute an important contribution to the marine scientific community. Being aware of the grave shortfall in the numbers of skilled people who

Figure 18.2 A formal gathering of all the world's hydrographers is held every 5 years in Monaco. © IHO.

can ensure the continuity of the GEBCO, a project to train a new generation of younger scientists and hydrographers in ocean bathymetry was submitted to the Nippon Foundation. This project started in 2004 and continues to run. Up to now over 30 trainees have reinforced personnel in hydrographic offices and oceanographic centres.

From a regional perspective, the development of comprehensive regional databases, which like their topographic equivalents for the land masses provide the underpinning charts or maps to support a wide variety of purposes including scientific research and modelling, exploration for natural resources, environmental policy making and marine education, has been a challenge for both the IHO and the IOC. The eight IBC projects have received and continue to receive strong support from the IHO through its Member States' Hydrographic Offices, as well as from other national, regional and international scientific institutions. SCOR and SCAR experts, as well as others engaged in marine scientific research, can make use of this detailed and accurate knowledge of the depth and shape of the seabed.

The monitoring of the variation of the sea level constitutes another area of great importance for the scientific community, which is searching for predictions of their change and effects over time. The IHO has an active Tidal and Water Level Working Group (TWLWG) that works closely with the well-established and supported Global Sea Level Observing System (GLOSS). For several years now these two groups have maintained close cooperation with representatives attending each other's meetings. We are seeing an increasing number of States appointing a common delegate to both groups thereby further enhancing collaboration. The IHO actively encourages its Member States to provide its tidal data both historically and in real-time to GLOSS. Sea level monitoring is a significant contribution to climate change studies.

International law now enables a Coastal State to claim additional maritime areas provided bathymetric and geophysical measurements support such claims. Article 76 of the United Nations Convention on Law of the Sea (UNCLOS) allows the Coastal State to exercise certain rights and assume specific responsibilities over their adjacent continental shelf even if it extends beyond the nominal 200 nautical mile EEZ. However, to claim such an extension, the Coastal State must delineate the extent of its continental shelf. Resulting from the new ocean regime established under the provisions of UNCLOS, the IHO established a specialized Working Group that in 1988 prepared the first edition of a Manual on Technical Aspects of the Law of the Sea (TALOS Manual – IHO publication S-51). In 1994 the TALOS Working Group became the Advisory Board on Hydrographic and Geodetic Aspects of the Law of the Sea (ABLOS). ABLOS was formed jointly with the International Association of Geodesy (IAG) and was established to continue the ongoing work of examining hydrographic, geodetic and other technical issues concerned with The Law of the Sea. As a measure to strengthen and broaden reciprocal cooperation the IOC joined the ABLOS in 1996 and participated actively with the IHO and IAG. They jointly prepared the current 4th edition of the TALOS Manual. Following the establishment of the IOC's Advisory Body of Experts on the Law of the Sea (ABE-LOS) charged with considering the full scope of UNCLOS in relation to the IOC, the IOC withdrew from ABLOS. This allowed the differentiation of the hydrographic and geodetic aspects of interest to the IHO and IAG from those of marine scientific research nature of interest to the IOC. Both the ABLOS and ABE-LOS groups continue to work in cooperation but independently to address their specific objectives.

The IHO recognized the importance of making bathymetric data available for scientific studies. IHO Member States are committed to making the necessary oceanic as well as coastal bathymetric data and information available. It does this through the IHO Data Centre for Digital Bathymetry (IHO-DCDB). Ocean bathymetry is extensively used for modelling tsunami travel time. Coastal data is vital for modelling the impact of the arrival of a tsunami and the determination of coastal run-up. The Global Ocean-related Hazards Warning and Mitigation System constitutes an important user of the IHO's effort in this sense. The IHO is fully aware of the fact that bathymetric information contained on nautical charts often represents only a relatively small subset of the total data available in national data centres within hydrographic offices. This data can contribute positively to the preparation of better analysis and modelling products such as inundation maps, a planning tool that could save many lives if available for each coastal community. Large-scale, precise hydrographic surveys using modern equipment are required to further improve such products but unfortunately not all coastal States have the capability to conduct such operations, and therefore the IHO assigns a high priority to capacity building.

The IHO also cooperates with the International Maritime Organization (IMO) and the World Meteorological Organization (WMO), as it does with the IOC, in the formulation of proposals for, and the execution of, technical projects that have

Figure 18.3 Traditional means of navigation were developed over centuries by different cultures. © Lagui/Shutterstock.com.

components of mutual interest that fall within the combined competence and expertise of the identified organizations. These projects have contributed to the enhancement of a global understanding of the importance of ocean sciences and have raised awareness on the need to exchange relevant information and adopt other measures to implement initiatives aimed at increasing the knowledge of the environmental conditions of the oceans and coastal areas. The Coast-Map Indian Ocean project is just an example of this joint partnership. Through this project the IHO and IOC have made best use of their respective strengths to transfer technology and know-how to 12 countries in the Indian Ocean.

For the IHO, Capacity Building is of strategic importance. The IHO has defined Capacity Building as '*the process by which the organization assesses and assists in sustainable development and improvement …*'. The capacity building strategy considers the four 'A's: Awareness; Assessment; Analysis; and Action. If governments do not assign a suitable priority to Ocean Sciences, including elements such as hydrography, the chances are that all efforts to improve existing capabilities will not be sustainable. The establishment of well-defined national institutions with clear mandates that are appropriately manned and funded have been the targets for the IOC and IHO during joint technical visits. Objectives of these technical visits have included: raising awareness of the issues; assessing actual capabilities; and identifying the processes and activities needed to develop oceanographic and hydrographic capability.

The IHO, as well as the IMO, IOC and WMO have together established mechanisms for the coordination of each organization's Capacity Building Program. This initiative was considered an excellent opportunity for these four intergovernmental international organizations to come together and cooperate and identify ways by which they could collectively improve capacity in areas of common interest. Through an analysis of the situation of each developing country visited, taken

Figure 18.4 Ocean shipping needs better charts for safer navigation. © Ilya Andrianov/ Shutterstock.com.

from the different perspectives of the IHO, IMO, IOC and WMO, more efficient and effective mechanisms have been found that lead to improved development of ocean sciences capabilities. Member States of all four organizations are aware of the importance of cooperation in the use of training facilities, research institutions, vessels, data and the expertise and experience of personnel, especially for the benefit of developing States.

To ensure best practice in coordinating our efforts, the IHO is represented in each other's main meetings, and vice versa. This, together with annual coordination meetings between the experts from the respective secretariats, ensures a continuously productive relationship that supports the achievement of our respective missions and objectives.

In brief, the IHO has been strongly contributing to the development and increase in visibility of the hydrographic component of ocean sciences. On the occasion of the International Year of the Ocean (1998) the United Nations General Assembly approved Resolution A53/32 where Article 21 invited States to carry out hydrographic surveys and to provide nautical services, making a clear reference to

Figure 18.5 Future power, but also a navigation hazard. Image by David Pugh.

Hydrographic Services. In November 2005 the United Nations through Resolution A60/30 welcomed the adoption by the International Hydrographic Organization of 'World Hydrography Day', 'to be celebrated annually on 21 June, with the aim of giving suitable publicity to its work at all levels and of increasing the coverage of hydrographic information on a global basis, and urges all States to work with that organization to promote safe navigation, especially in the areas of international navigation, ports and where there are vulnerable or protected marine areas'. Now the IHO stands ready to consider UN Resolution A63/111, which adopted 'World Oceans Day' that was celebrated for the first time on 8 June 2009. The IHO is conscious that these celebrations offer a window of opportunity to make society aware of the importance of all ocean issues and to consider the actions needed to undertake national and individual responsibilities to sustain the Oceans – the greatest common heritage that we have and without which we cannot exist.

Decision-making success depends on timely, adequate and reliable information. Hydrography is one of the relevant and basic guiding elements to be considered in almost all uses we make of the seas and oceans. The IHO looks forward to further joint work with the IMO, IOC and WMO and with other scientific bodies, contributing to safety of life at sea, safety of navigation, protection of the marine environment and contributing to the ocean sciences as a means to having a better understanding of our planet.

Descriptions of mission, structure, components and work of the IHO can be found at http://www.iho.org.

19 Ocean science and shipping: IMO's contribution

RENÉ COENEN

René Coenen, an economist by training, has worked throughout his career on marine environmental protection issues in and outside his native Netherlands, in particular the OSPAR Commission, the London Convention, North Sea Ministerial Conferences, GESAMP and the North Sea Working Group. He joined the IMO in 1992, where he is now Deputy Director, and Head of the Office for the London Convention and Protocol.

The Convention establishing the International Maritime Organization (IMO) was adopted in Geneva in 1948 and the IMO first met in 1959. As a Specialized Agency of the United Nations with 169 Member States, the main task for the organization is to develop and maintain a comprehensive regulatory framework for shipping. This mandate includes maritime safety, environmental concerns, legal matters, technical cooperation and capacity building within national administrations, maritime security and the efficiency of shipping – as summed up in its vision of: 'Safe, Secure and Efficient Shipping on Clean Oceans'.

Having, in the first decade of its existence, given priority to the development of treaties dealing with maritime safety, the IMO increasingly turned its attention to marine pollution issues, a concern emerging in the late 1960s. The growth in the amount of oil transported by sea and in the size of oil tankers was of particular concern and the Torrey Canyon disaster in 1967, in which 120 000 tonnes of oil were spilled, demonstrated the scale of the problem.

While preparing for the 1973 Marine Pollution Conference in London, which resulted in the adoption of the International Convention for the Prevention of Pollution from Ships (the MARPOL Convention), the IMO realized that its hitherto technical and regulatory work needed a solid basis in marine science. Initially, the IMO drew directly, on an ad hoc basis, on the scientific community through its membership, but it realized that more structured arrangements for involving marine scientific advice were required and that the

Figure 19.1 Littleton Harbour, Christchurch, New Zealand. Ports and sea transport are the lifeblood of the global economy. Ports are surrounded with areas of industrial activity. Image by David Pugh.

IMO was not equipped to set these up on its own. This resulted in closer cooperation with intergovernmental and international science organizations, such as the IOC.

In the 1980s, the IOC made a major contribution to the IMO's work in developing methodologies for sampling and analysis of petroleum hydrocarbons and combustion products in seawater, in biota, in sediments and in the sea-surface micro layer; developing the possibility of identifying the origin of the oil detected and developing initial assessments of observed oil levels in different regions of the world (in particular, the West and Central African region; the Mediterranean Sea; the ROPME Sea area (Regional Organization for the Protection of the Marine Environment (Kuwait); and the south-east Pacific Ocean). As most of the available data relating to dissolved/dispersed petroleum hydrocarbons, at that time, were from near-shore, coastal and estuarine zones and did not show conditions in open-ocean areas, the IOC's offer of further cooperation in this respect was welcomed by the IMO.

In the 1980s, the IOC, IMO and UNEP jointly organized a workshop on 'biological effects' measurements with a view, *inter alia*, to establishing the scientific basis for the definition of vulnerability of marine areas to marine pollutants. The IMO recognized the value of these developments, in the context of the identification of particularly sensitive sea areas, a requirement identified by the International Conference on Tanker Safety and Pollution Prevention in 1978. This, and similar initiatives, greatly contributed to the development, within the IMO, of the 'Particularly Sensitive Sea Area' (PSSA) concept and culminated in the adoption of 'Guidelines for the Designation of Special Areas and the Identification of Particularly Sensitive Sea Areas' by IMO Assembly resolution A.720(17) in 1991 (the PSSA Guidelines).

The PSSA Guidelines were designed to assist in providing guidance to IMO Member Governments in the formulation and submission of applications for PSSA designation. Three elements must be present for identification and designation as a PSSA: first, the area must have certain specific attributes (ecological, socio-economic or scientific); second, it must be vulnerable to damage by international shipping activities; and third, there must be associated protective measures with an identified legal basis that can be adopted by the IMO to prevent, reduce or eliminate risks from these activities. An area designated as an internationally recognized PSSA by the IMO will have one or more IMO-adopted associated protective measures for ships to follow. In 2005, the IMO finalized a comprehensive revision of the PSSA Guidelines and adopted these by Assembly resolution A.982(24). As a result, to date, 12 PSSAs have been designated worldwide.

Within the GIPME framework, IOC, UNEP and IMO collaboration over a period of several years resulted in the adoption of a number of measures, notably, in the period 1997 to 2000, the development of approaches for evaluating the degree to which contaminated marine sediments might adversely affect marine resources. Several attempts to develop these approaches had been made before that time but none had been widely accepted. The collaboration resulted in the publication, in 2000, of the 'GIPME Guidance on Assessment of Sediment Quality', which examined various approaches to assessing anthropogenic impacts on marine sediments, and associated risks to marine life and human health.

The conclusion in that report to the effect that numerical sediment quality criteria are unsuitable for widespread application, and that the scientific basis for assessing sediment quality must incorporate biological, chemical and physical considerations, has since been incorporated in the 'Specific Guidelines for Assessment of Dredged Material', which were adopted in 2000 by Parties to the Convention on the Prevention of Marine Pollution by Dumping of Wastes and Other Matter (London Convention 1972). These guidelines provide advice on how to assess the quality of the 200 to 400 million tonnes of dredged material disposed at sea annually, under licence.

Since the inception of GESAMP in 1969, the IOC, IMO and its other sponsoring organizations have collaborated closely on many issues of shared concern. GESAMP's annual sessions are hosted on a rotational basis. GESAMP Working Groups prepared many scientific advisory papers on the marine environment, including important independent reviews of the health of the ocean. GESAMP celebrated its 40th anniversary in 2009 and the collaboration between its sponsoring organizations continues to flourish, providing authoritative, independent, interdisciplinary scientific advice to organizations and Governments to support the protection and sustainable use of the marine environment.

In the 1990s, a 5-day introduction course to ocean sciences and marine pollution management was given, under the IOC's supervision, as part of the curriculum of the World Maritime University (WMU) in Malmö, Sweden. This was done in

Figure 19.2 The main task for the IMO is to develop and maintain a comprehensive regulatory framework for shipping. This mandate includes environmental concerns. An engineer is seen taking samples from mudflats adjacent to a port. © Keith A. Frith/Shutterstock.com.

recognition of the fact that WMU graduates from developing countries need a basic understanding of these issues when employed in the maritime sector.

Addressing the problem of the introduction of harmful aquatic organisms and pathogens to new environments through ships' ballast water (which has been identified as one of the four greatest threats to the world's oceans), led to the adoption, in 2004, of the IMO Ballast Water Management Convention. Proper control and management of ships' ballast water poses a major and complex environmental challenge for the IMO and the global shipping industry and depends, to a large extent, on the ability of Parties to the Convention to develop marine environmental baselines and to use other scientific tools to monitor the effectiveness of the implementation of this Convention.

Similarly, the current efforts of the IMO to develop measures for the reduction of greenhouse gas (GHG) emissions from international shipping, as its contribution to the overall GHG reduction targets the world community is setting as a follow-up to the Kyoto Protocol, rely heavily on ocean sciences and their ability to monitor the effectiveness of such measures.

Figure 19.3 IMO building - Seafarers Memorial (HQ of the International Maritime Organization of the United Nations in London), and the Houses of Parliament in the background. © R.L. Pulido, http://www.flickr.com

Finally, in 2008, Parties to the London Convention and Protocol adopted a non-binding resolution (LC-LP.1 (2008)) on the regulation of ocean fertilization. By this resolution, Parties have declared, *inter alia*, that, 'given the present state of knowledge, ocean fertilization activities other than legitimate scientific research should not be allowed'. In so doing, they acted on the 'Statement of Concern' of their Scientific Groups, in June 2007, that knowledge of the effectiveness and potential environmental impacts of ocean fertilization was currently insufficient to justify large-scale operations and that these could have negative impacts on the marine environment and human health. Further work on a potential legally binding resolution, or an amendment to the London Protocol on ocean fertilization, is to follow in 2010, together with a document summarizing the current state of knowledge on ocean fertilization, relevant to assessing impacts on the marine environment, taking into account the work done on this issue in other fora (e.g. the Biodiversity Convention). To enable well-informed decisions in these fora, the IOC has coordinated the preparation of a 'Summary for Policymakers for Ocean Fertilization'.

The above three examples demonstrate the cooperation required between regulatory systems and the knowledge and information on which they must be based. Future environmental agreements and regulations will continue the need to have a strong basis in marine sciences. This, in itself, gives the potential for continued and fruitful cooperation with the international and intergovernmental science community.

20 The UNEP's contribution to the oceans and marine science

SALIF DIOP[1]

Salif Diop is a Senior Programme Officer at the UNEP's Division of Early Warning and Assessment (DEWA). He is a water specialist with extensive experience in various aspects of coastal oceanography, freshwater assessment, aquatic and marine issues, sustainable management and development.

After almost 40 years, the sustainable management of marine and coastal resources remains a significant challenge for the United Nations Environment Programme (UNEP) and the global community, particularly in view of the international commitment to the Millennium Development Goals in 2000.

The UNEP has continually adapted its approach to marine and coastal management to ensure delivery on its mandates. Recent reviews of the UN system as a whole called for greater coherence across various development-related agencies, funds and programmes. The UN Secretary-General noted, in particular, the need to strengthen delivery of operational activities. Cognizant of recent directional shifts in the organization, the Medium Term Strategy (MTS) 2010–2013 called for a coordinated, results-focused delivery of work programmes. Six cross-cutting priorities were identified to provide a new focus for the work of the UNEP, namely: climate change; disasters and conflicts; ecosystems management; environmental governance; harmful substances and hazardous waste; and resource efficiency.

The UNEP identified within its new marine and coastal strategy assessment/ knowledge, management, policy and partnership gaps and the UNEP's comparative

[1] Prepared in cooperation with Jacqueline Alder, Head of Coastal and Marine Programme.

Fig 20.1 UNEP has a mandate that encompasses the marine environment and all its living creatures. © Edwin van Wier/Shutterstock.com.

advantage in addressing specific marine issues, and identified seven priority areas in the marine and coastal sector:

1. Pollution from land-based activities (LBA) including excessive nutrients.
2. Physical Alteration and Destruction of Habitats, including through aquaculture.
3. Impact of climate change on oceans and coasts.
4. Marine and coastal biodiversity, including deep seas.
5. Environmental aspects of fisheries.
6. Environmental aspects of high seas and seabed management and governance (beyond areas of national jurisdiction).
7. Vulnerability of islands.

The preceding directional shifts within the UNEP have provided the management driver for the UNEP to commit to the development of a dedicated Marine and Coastal Strategy within the Ecosystem Management theme but to cut across the six themes and guide future work in this area. The present strategy uses a 10–15-year horizon for facilitating changes in the marine sector, while recognizing and complementing existing strategies, plans and policies including MTS 2010–2013, the UNEP Water Policy and the Climate Change Strategy. It aims to build on the identified strengths

Fig 20.2 Marlborough Sound, Queen Charlotte Islands. Establishing Marine Protected Areas is a difficult but important goal for Coastal States. Taking the next step into international waters will be even more challenging. © Holger Mette/Shutterstock.com.

and experiences of the UNEP as an organization, to facilitate change at the policy and operational levels.

Strategy initiatives are grouped in four key streams, each addressing a number of the priorities and expected accomplishments identified in the 2010–2013 strategy and beyond. The four streams of activity are:

- Land–Coast Connections
- Marine and Coastal Ecosystems for Humanity
- Reconciling Resource Use and Marine Conservation
- Vulnerable People and Places

Since the UNEP was established, the Regional Seas Programme (RSP) has been one of its flagship programmes. The RSP gives priority to regional activities, encouraging and supporting the preparation of regional agreements for the protection of specific water bodies. To date, UNEP has supported the negotiation, adoption and implementation of 13 regional seas conventions and action plans throughout the world, and become the Secretariat of six of these, as well as created partnerships with five marine programmes. The UNEP provides ongoing support to regional seas governing bodies on legal and public relations issues, and assists in the achievement of financial sustainability. The regional seas conventions and action plans are also used as instruments for sustainable development and as platforms for regional implementation of multilateral environmental agreements, programmes and protocols. The UNEP and IOC of UNESCO have been active partners in using the Regional Seas platforms.

Following various calls by the UNEP Governing Council and relevant preparatory work, the international community adopted the Global Programme of Action for the Protection of the Marine Environment from Land Based Activities (GPA) in 1995, addressing the interlinkages between the freshwater and coastal environments. The UNEP was called to support States in implementing sustained action to prevent, reduce, control and/or eliminate marine degradation from land-based activities through applying integrated coastal area and watershed management approaches.

At the request of governments, the UNEP became the GPA Secretariat in 1997. Further to the UN reform process being carried out in those years, the nineteenth session of UNEP GC in 1997 adopted the Nairobi Declaration on the Role and Mandate of the UNEP.

In Agenda 21, adopted at the Rio Conference on Environment and Development, states committed themselves to improve the understanding of the marine environment in order to assess present and future conditions more effectively (UN 1992). In 2001–2002, work commenced to explore the feasibility of establishing a regular global assessment process for the state of the marine environment. The feasibility study led the 2002 World Summit on Sustainable Development in Johannesburg to support actions at all levels to '*establish by 2004 a Regular Process under the United Nations for global reporting and assessment of the state of the marine environment, including socio-economic aspects, both current and foreseeable, building on existing regional assessments*'. In November 2005, UNGA launched the 'Assessment of Assessments' (AoA) as a preparatory stage towards the establishment of the Regular Process (resolution 60/30). An Ad Hoc Steering Group (AHSG) was established to oversee implementation of the AoA and a Group of Experts was established to carry it out, supported by a secretariat of the two lead agencies: UNEP and IOC of UNESCO. These agencies have, since 2006, proceeded with the implementation of the start-up phase of the Regular Process, producing an excellent partnership in complementing each other in their respective comparative advantage in marine environmental science and assessment. The vital role of the UNEP in undertaking assessments, such as the Global Earth Observation System of Systems (GEOSS) process and the UN Assessment of Assessments, provides critical information and analysis on conditions and trends in marine and coastal environments.

Where does the Global Reporting and Assessment of the Marine Environment (GRAME) fit in the world's oceans needs at this time?

The marine environment is undergoing unprecedented change, threatening ecosystem goods and services on which humans depend. The response from the international community and national governments involves judgements about trade-offs among objectives and priorities, and about the likely effectiveness of policy options. These judgements will need to be based on up-to-date information and supported by iterative assessments. To promote support for such policy-making, the United Nations General Assembly (UNGA) decided to launch a start-up phase towards a regular process for global reporting and assessment of the state of the marine environment, including socio-economic aspects. After presenting the report – 'Assessment of Assessments' – on the outcome of this preparatory stage, the UNEP and IOC UNESCO hope that UN Member States will be empowered with information on the framework and options for a future global regular process as well as a valuable resource for decision-makers dealing with marine environmental issues.

21 The World Meteorological Organization need for ocean science

PETER DEXTER AND YVES TREGLOS

Peter Dexter was trained as a meteorologist and worked as a weather forecaster for a short while before returning to university to take his masters and doctorate in ocean remote sensing using HF radars. He then undertook research for the Australian Bureau of Meteorology in ocean wave and storm surge modelling and air–sea interaction, before moving to the World Meteorological Organization in Geneva in 1984.

Yves Tréglos prepared the first objective analysis of sea surface temperature in the Bay of Biscay. He then entered CNEXO, the predecessor of IFREMER. He was elected as the first chairman of the Joint IOC-WMO Committee for IGOSS. He was recruited by WMO as Scientific Officer, seconded to the IOC. He served in that linking capacity up to his retirement, in 2003.

The World Meteorological Organization (WMO) is a specialized agency of the United Nations. It is the UN system's authoritative voice on the state and behaviour of the Earth's atmosphere, its interaction with the oceans, the climate it produces and the resulting distribution of water resources. The WMO has a membership of 188 Member States and Territories. It originated from the International Meteorological Organization (IMO), which was founded in 1873. Established in 1950, the WMO became the specialized agency of the United Nations in 1951 for meteorology (weather and climate), operational hydrology and related geophysical sciences.

Figure 21.1 Monitoring of the tropical atmosphere and ocean with the TOGA/TAO array. © NOAA, courtesy of JCOMM.

Meteorology and the mariner

Humanity has long been simultaneously fascinated and awed by the powers of the air and sea, as well as anxious to understand and exploit the processes observed. Lacking anything beyond a basic empirical knowledge, the early seafarers remained at the mercy of wind, waves and currents, and whatever was driving them. Fortunately, mariners, to survive and prosper, have to be intensely practical people. In extending their trading and exploration voyages over wider and wider sea areas, and to more and more distant lands, they accumulated a formidable body of empirical knowledge of the atmospheric and oceanic environment in which they lived and worked. This knowledge was a sound basis for advances in scientific understanding, and the development of predictive capabilities.

The international cooperation that was set in train by the Brussels Maritime Conference of 1853 led directly to the First International Meteorological Congress in Vienna in 1873, and ultimately to the formal establishment of the IMO in 1905. In parallel with these international developments, several countries were in the process of establishing their own national meteorological agencies during the years 1850 to 1870. As in so much of early meteorology, this was stimulated by the needs of the maritime community. Thus began the modern era of meteorological service interaction with and support for this community.

The transition from sail to steam at the end of the nineteenth century, and the consequent belief that the safety of maritime transport might gradually become less critically dependent on meteorological information, together with the advent of aviation as the primary focus for meteorological services in the first half of the twentieth century, resulted in some loosening of the traditional close ties between meteorology and the mariner. However, recent years have seen a reversal in this trend, with a number of factors each playing a part: the recognition that a large majority of maritime safety incidents (up to 70%) remain weather-related; new communications

Figure 21.2 A young boy and other fishermen catching fish in a field inundated by a tidal surge in Chittagong, Bangladesh. © Jashim Salam/Marine Photobank.

technologies, which allow the reliable delivery of an enhanced range of maritime safety information to ships at sea; the development of much more specialized shipping, requiring a balance among safety, minimizing the potential for cargo damage, reducing voyage times and fuel costs, and managing increasingly busy ports and seaways; and the opening up of new sea routes, especially in polar waters. All these factors are contributing to a new recognition of and reliance on the delivery of high-quality meteorological and oceanographic information to ships at sea.

Early needs for ocean data and science: the establishment of IGOSS

The 1960s saw a strengthening appreciation within the global meteorological community of the important role played by the oceans in atmospheric processes, in determining fluxes of heat, momentum and moisture, to and from the atmosphere. The advent of the World Weather Watch, and of numerical weather prediction, brought a requirement to at least parameterize air–sea exchanges, and to measure and transmit some basic ocean variables in near real-time, notably sea surface temperatures. Thus the interests of the WMO and National Meteorological

Services were rapidly extending beyond the traditional sea surface winds and waves to more fundamental ocean variables. At the same time, a small number of far-sighted people in the global oceanographic community, represented by the IOC, recognized a growing requirement to more systematically sample and analyze ocean processes, with a view to delivering ocean data, products and services to various marine users.

The IOC, with its recognized competence in ocean sciences, and WMO, with its recognized competence in operational meteorology, needed to cooperate in the field of air–sea interactions, and would derive mutual benefit from this cooperation. The idea eventually materialized in 1967, through the establishment of a joint panel of experts, which immediately recommended that both organizations '*take the necessary steps to set up an integrated global ocean data system serving both meteorology and oceanography, which ensures that all countries obtain the global ocean observational data they require*'.

Such wording was familiar to meteorologists. At that time, however, it sounded revolutionary to many oceanographers. The old joke: 'There are two kinds of data, my data and bad data', was still with us. Regarding ocean data, scientists were fonder of quality than of quantity (they had some rationale for that). Oceanography was a science, and the wording 'operational (or even synoptic) oceanography' was often considered as rude if not meaningless.

Nevertheless, times were about to change. IOC and WMO are *intergovernmental* (as opposed to international) scientific organizations. The issue was not to provide just scientists, but *countries*, with the data they required. The aim was not solely to increase our understanding of the planet, but to serve practical ends that contributed to analyzing, before hopefully predicting, some environmental features that were of concern to humans. In this framework, science was no longer considered as the target, but rather as a tool. That implied a considerable change in mentality, especially for oceanographers.

The IOC and WMO regarded establishing an Integrated Global Ocean Services (initially *Station*) System (IGOSS) as the way forward. IGOSS was conceived as consisting of national facilities and services and benefiting from intergovernmental coordination and support. It was aimed at being responsive to the operational and research requirements agreed upon among the participating nations. In addition, it had, *inter alia*, two characteristics:

1. everything within the system was supposed to be *fully standardized and uniform*;
2. the top priority was to develop *international product generation services*.

IGOSS was the first attempt to establish an embryo of operational oceanography worldwide. It critically reviewed its contribution to the First GARP Global Experiment (FGGE) in the late 1970s and endeavoured to improve its functioning, notwithstanding criticisms from 'traditional' oceanographers. While in some respects, IGOSS was an idea before its time, today the basic idea is still alive, at a much

Figure 21.3 A ship is threatened by an approaching tornado. © Dark O/Shutterstock.com.

larger scale, within the Joint Technical Commission for Oceanography and Marine Meteorology (JCOMM).

The oceans and climate

The idea of an international research programme on climate and climate change developed out of the Global Atmospheric Research Programme of the 1970s, it was given impetus by the first World Climate Conference, and the WCRP was formally established by the WMO in 1980, co-sponsored by the ICSU. The scientific plan for the WCRP clearly recognized that the oceans, land surfaces, the cryosphere and biomass all needed to be taken into account and incorporated into global climate models. The critical role of the oceans in particular in the climate system meant that close cooperation was essential between the meteorological and oceanographic communities, and the IOC joined the WMO and ICSU in sponsoring the WCRP in 1993. Two major experiments of the first two decades of the WCRP were seminal in establishing the importance of the oceans to WMO, at least in the context of climate science:

1. The Tropical Ocean Global Atmosphere project (TOGA, 1984–1995) clearly demonstrated an essential requirement for meteorologists and oceanographers to work together in addressing climate science and prediction; it also demonstrated to the oceanographic community the value of transmitting, processing and disseminating oceanographic observational data in near real-time.
2. The World Ocean Circulation Experiment (WOCE, 1990–2002), the most ambitious oceanographic experiment undertaken to date, aimed to clearly establish the role of the oceans in the climate system, and to obtain a baseline dataset for assessing future change; it also contributed directly to the development of the Argo profiling float, and to the later Global Ocean Data Assimilation Experiment (GODAE).

Both TOGA and WOCE firmly established in WMO the importance of systematic, long-term ocean monitoring, of transmitting and disseminating the data in real-time, and of assimilating these data into ocean, and coupled ocean–atmosphere, models to understand and predict climate and climate variability. This was a big step towards engaging the global meteorological community in the concept of operational oceanography.

The realization of operational oceanography

The experience gained and lessons learned through IGOSS, coupled with the undoubted success of TOGA and WOCE in demonstrating that the oceans could be monitored systematically, using new, innovative and cost-effective *in situ* technologies, with quality data delivered on timescales appropriate to real-time applications, and backed by new generations of ocean satellites and super-computers, had paved the way towards operational oceanography. Nevertheless, obstacles remained, in particular in the lack of relevant national institutional arrangements or even cooperation, together with a plethora of international bodies poorly adapted to the new requirements. In addition, a clear demonstration was required that the new technologies, new understanding of ocean processes and new support facilities could be successfully combined to deliver ocean analysis and forecast products of value to society. Fortunately, there were sufficient people available, at the appropriate time, in the international oceanographic and meteorological communities to overcome the obstacles and realize the potential.

Following a remarkably short and effective gestation period, the 13th WMO Congress and the 20th IOC Assembly, in 1999, approved the establishment of the JCOMM. The JCOMM was expected to address, *inter alia*:

Figure 21.4 Storms over a rough sea. © Studio 37/ Shutterstock.com.

- the pressing need for a fully coordinated joint mechanism for implementing the stated requirements for ocean and surface marine meteorological data to support GOOS/ GCOS …
- the expanding requirements of many other marine users for a comprehensive range of both meteorological and oceanographic data and products
- the potential benefits to be gained from making better use of the diverse and extensive range of expertise and facilities available to both organizations …

How well these expectations have been fulfilled will be for history to judge, but the initial indications are positive.

In 1997, acknowledging the need for better ocean observations and ocean forecasts and with the scientific and technical opportunity that readily available satellite data had delivered, the Global Ocean Data Assimilation Experiment (GODAE) was initiated, to lead the way in establishing global operational oceanography. The vision was for a *global system of observations, communications, modelling and assimilation, that will deliver regular, comprehensive information on the state of the oceans, in a way that will promote and engender wide utility and availability of this resource for maximum benefit to the community*. The GODAE as an experiment finished in 2008, with the follow-on activities now closely associated with the JCOMM, as the appropriate mechanism to carry forward the coordination of its results at the intergovernmental level. More importantly, ocean analyses and forecasts are now being delivered routinely to users in several parts of the world. While the skill levels may not yet match those of operational meteorology, it is fair to say that the potential is finally being realized.

WMO and IOC: a marriage made in heaven?

Two concepts have always been fundamental to the meteorological world, and to the WMO: meteorological science, rather than an end in itself, is a tool to better understanding atmospheric processes and ultimately to delivering products and services of direct value to society; and the free and open exchange of data, as rapidly as possible, is critical to the delivery of such services. Until recent years such concepts have been somewhat alien to oceanographers. To a certain extent, this has coloured the relationship between the meteorological and oceanographic communities, both nationally and internationally, over several decades. When coupled with a fear in the ocean community (not entirely irrational) of international meteorology as 'big brother', this led to some degree of friction between the WMO and IOC during the 1960s and 1970s. Fortunately, times change, and people change, and the 1980s and 1990s saw a significant improvement, based on an increasing range of interdependencies, enhanced mutual understanding and personal confidence.

There is no doubt that the dependencies and commonalities for the two communities are growing: operational meteorology and numerical weather prediction (NWP) increasingly require dynamically coupled atmosphere/ocean models and at least full ocean mixed layer analyses, while ocean biogeochemical processes and feedbacks are a significant factor in the global climate system, and a component of the shift to whole earth system science. At the same time, atmospheric and climate processes and variability have a profound influence on ocean biology and chemistry, and all branches of marine science need to understand and utilize these interactions. Additionally, it makes practical and economic sense for the ocean community to utilize, to the extent possible, the existing facilities and infrastructure of operational meteorology. And finally, the provision of data, products and services to deliver societal benefits, including in particular preparedness for, warning of and response to a range of natural hazards, demands that the two communities, from the intergovernmental level through to local actions, work ever more closely together.

Of course, sceptics remain on both sides: many National Meteorological Services (NMS), reflected in WMO policies, remain unconvinced that oceanography is important to their core business; while in the IOC, the continuing predominance of representatives from marine research institutions and/or Government policy-making organs leads to an ongoing reluctance to embrace operational concepts and their associated levels of cooperation and free and open data exchange. Nevertheless, while the relationship of WMO and IOC may not be a marriage, nor made in heaven, it surely will involve cohabitation, and a reflection of practical, earthly concerns.

Part VII The Future: Preface

The next half-century

Our journey through the last 50 years in the development of ocean science, and its application by governments to the management of the oceans, has been both informing and intriguing. Several authors involved in that evolution have given informative accounts from their individual perspectives: several have also given us a snapshot of where we are today. An intriguing aspect of this is whether their narrative and analysis of that process equips us to look ahead and predict events over the next 50 years and beyond.

The answer to that question must be ambivalent. Correct forecasting can make fortunes, but few succeed to become wealthy in this way! Writing in 1964, and looking ahead to Orwell's portentous *1984*, in an article for the *New Scientist*, Roger Revelle described the view from a beach in California. In 1984, he predicted that hurricanes are controlled and eliminated by covering vast areas of ocean with clouds of heat-absorbing aluminium oxide; cooling waters from nuclear power stations are used to create year-round ideal sea temperatures for bathing, so encouraging the spread of mega coastal resort cities; fresh water is towed from river sources to dry areas in 10 km long plastic submerged containers; wealth from the mining of manganese nodules has funded the UN system; and a global ocean observing system allows accurate long range weather forecasting. Even today, in 2010, only the last of these predictions is becoming a reality. If someone with Revelle's vision misses so many targets, then the rest of us must indeed proceed with caution.

We have invited Neville Smith to write a forward-looking chapter. As he explains, the context of the 50th anniversary of the IOC provides an opportunity for us to take stock, to reset and to energize the role of science in ocean management for the next 50 years. Finally, as editors, and with strong encouragement from our publishers, we have attempted to take an overview of the issues discussed in the previous chapters. The chapters contained in this book do not represent a comprehensive reference work on the subject of Ocean Science and Governance; however, collectively, they contain the thoughts of many experienced and knowledgeable people and, as such, provide a basis upon which the opinion of the reader can be developed. These two final contributions are intended as a stimulus for further debate and perhaps as a checklist against progress at some future date.

22 The future

NEVILLE SMITH

Neville Smith, a leading scientist with the Australian Bureau of Meteorology, has played a key role in the development of ocean and climate observations and prediction systems during the last two decades, nationally and internationally. He led the Ocean Observations Panel for Climate for 5 years, when he initiated and led the Global Ocean Data Assimilation Experiment. He was Vice-Chair of the Intergovernmental Oceanographic Commission from 2005 to 2009.

Introduction

This volume is dedicated to the theme of Ocean Science and Governance. The broader context provides an opportunity for us to take stock, to reset and to energize the role of science in ocean management for the next 50 years. Here the scope and mandate of the present Intergovernmental Oceanographic Commission, whose 50th anniversary this book acknowledges, will frame my view of the future. The IOC was established and remains the principal mechanism for intergovernmental cooperation in ocean science. I shall not attempt to cover all issues that might be relevant to the future of the oceans: my perspective will be eclectic rather than comprehensive.

How far into the future might one dare look? Is a 50-year perspective realistic? For reference, major decisions of the Commission (for example, data policy) typically take around 4 years before conclusion; trends in science usually manifest over 3–5 years; ocean technology usually takes up to 10 years from idea to operations; conventions or the creation/modification of intergovernmental mechanisms can take longer than a decade; and influencing the evolution of climate change takes 50 years or more: the next 30 years of change is already locked in.

The marine science and technology that will be the focus of the Commission in 2020 is probably already foreseen but little of what will be prominent 50 years hence can be known. Major intergovernmental, societal, environmental and economic

Figure 22.1 Corals can be encouraged to grow around structures to create artificial reefs. © NOAA.

changes during the next 50 years may occur, but the nature of that change is largely unpredictable. Given this context, the discussion will mostly focus around the next decade or so, but even then there is much uncertainty. The excitement generated by new discoveries, improved understanding and revolutions in technology will be tempered by the threats arising from natural disasters, climate change and the inevitable pressures placed on the marine environment and its ecosystems.

In a formal sense, the future of the IOC in this development was examined through an open Working Group reporting to the Executive Council in 2008. The Working Group discussed a wide variety of issues including the mandate of the IOC, its operation and Member State involvement. The conclusions of that Working Group will be used to develop an image of the IOC, and indirectly of intergovernmental marine science cooperation more generally, in 2050.

Although it is usual in exercises such as this to dwell on the exciting, on the aspirations, on the dreams and take a generally optimistic view of the future, such a perspective might not be a realistic analysis. For a number of reasons, some of the reflections below will appear sober, perhaps even pessimistic, but hopefully still useful and interesting.

Learning from the past

Earlier chapters in this volume provide an account of 50 years of ocean science and governance. In some of these the role of the IOC is incidental, in others fundamental. The changes in science, in technology, in the geopolitical and socio-economic environment, and in the intergovernmental system itself have been huge and profound. Since past trends are always a good indicator of future trends, it is useful to devote a little time to those aspects that have most deeply shaped the situation as we know it today.

Figure 22.2 This nineteenth century pier was built for holidaymakers, from unwanted railway lines. How versatile will we be in our twenty-first century use of the oceans? Image by David Pugh.

Transformational events during the past 50 years

At the time of the creation of the IOC there was little appreciation of the mesoscale variability which we now know dominates the ocean; there were no numerical models of the ocean; and we were still several years away from seeing the first satellite images of the ocean. Our knowledge of the global ocean was scant and restricted predominantly to the North Atlantic and North Pacific, and there was but the barest appreciation of coastal ecosystems and marine biodiversity. Global ocean experiments were still several decades away.

The 1980s was the period when ocean science, and its organization, changed fundamentally. Scientific understanding of the time suggested that El Niño, a period of elevated warm temperature in the eastern tropical Pacific that greatly reduced anchovy harvests, was connected to a Pacific-wide change in both the oceanic and atmospheric circulation, probably through large-scale ocean–atmosphere feedbacks. The community realized that progress with understanding and predicting El Niño would not emerge from isolated individual pursuit but through internationally coordinated observation, modelling and process studies. Through 1985–1994, the Tropical Oceans-Global Atmosphere (TOGA) programme aimed and succeeded in developing useful climate prediction systems, with the legacy of that achievement remaining in the form of an array of tropical ocean observing networks and a number of model-based El Niño prediction systems.

Among its many achievements, TOGA developed the on-going community of scientists and ultimately practitioners in climate observation, analysis and prediction that is needed to tackle global-scale, multinational problems. Such communities are fundamental to the way modern intergovernmental mechanisms work, operating at a scale that fits intergovernmental procedures.

Perhaps the single most important TOGA legacy was a system of real-time, open and free exchange of scientific data, building on the early steps taken by the sea level community. The current state of the tropical ocean is known in real-time and any nation with risks related to El Niño can develop warning systems based on an observational and knowledge base that is now free and open. While TOGA had strong roots in the meteorological community, the World Ocean Circulation Experiment (WOCE), which started in the early 1990s, can be viewed predominantly as an initiative of the ocean science community. The WOCE was pioneering in its extent, in the technologies it introduced, and in its detailed modelling of the ocean. The WOCE observations provided the reference against which climate changes in the ocean are studied today. Through altimetry, *in situ* sea level measurements, and a range of other observational and modelling approaches the WOCE changed the way we do ocean science.

The community that the IOC helped bring together for the WOCE were vital in seeing the TOPEX-POSEIDON altimeter mission realized, which has provided unprecedented global measurements of sea surface elevation variability and change. Altimetry has now become a core contribution, albeit with some work still to do to ensure its long-term sustainability. Another technological innovation, the profiling float, led to a revolution in the way we observe the upper ocean. In its short 10-year life it has already contributed fundamentally to improved understanding of climate change in the oceans and constitutes the central building block upon which operational ocean observing systems are being constructed.

Together with the WOCE's major impact on modelling, the above examples illustrate how ocean science has changed and evolved during the past 50 years. Other experiments such as the Global Ocean Data Assimilation Experiment (GODAE), Global Ocean Ecosystems Dynamics (GLOBEC) and the Joint Global Ocean Fluxes Study (JGOFS) could be used to illustrate some of these points but it was the transformational role of the TOGA and WOCE that was fundamental. In 20 years the way the oceans were observed and modelled, and the way data was communicated and made available, changed fundamentally, paving the way for a new approach. The same evolution also influenced the character of the challenges for the IOC, particularly in terms of governance for data collection and exchange.

Evolution of ocean science governance

As the major intergovernmental body for ocean science and services, the IOC of UNESCO has a number of functions. During the last 50 years, its major function was in terms of enabling global scale science that might otherwise not have been possible. While few among the ocean science community would single out intergovernmental mechanisms as critical for the success of global ocean experiments, the backdrop of governmental agreement and cooperation is and

Figure 22.3 Travelling among the Solomon Islands. The future of the oceans is unclear! Image by David Pugh.

remains an important enabler, for initiation and for sustaining the legacy of the experiments. The IOC also contributed directly to science through ocean observing initiatives but, in terms of assessing trends for the future, it is probably the intergovernmental fabric that will be more fundamental and enduring.

The IOC is neither a science programme nor a science funding body. The IOC coordinates and facilitates actions by its Member States, seeking cooperation and joint activities for the benefit of the majority rather than the few. Participation has grown and the centres of influence have evolved, with countries like China, India and Brazil being increasingly influential. The IOC permits government-to-government cooperation but the underlying actions must be supported by the Member States themselves.

There is no Convention underpinning the operation of the IOC, so decisions are not binding or enforceable. Obligations manifest as best endeavours, not mandatory undertakings. For example, the IOC data policy is used as a guide and an inducement, rather than a mandatory requirement for activities under the IOC banner. The demands on the IOC are immense, and in keeping with the growth in importance and complexity of ocean issues, yet its actions are rarely able to meet expectations. Member States express many expectations from the IOC and yet are simply not willing or able to channel the necessary resources into the IOC that such expectations demand.

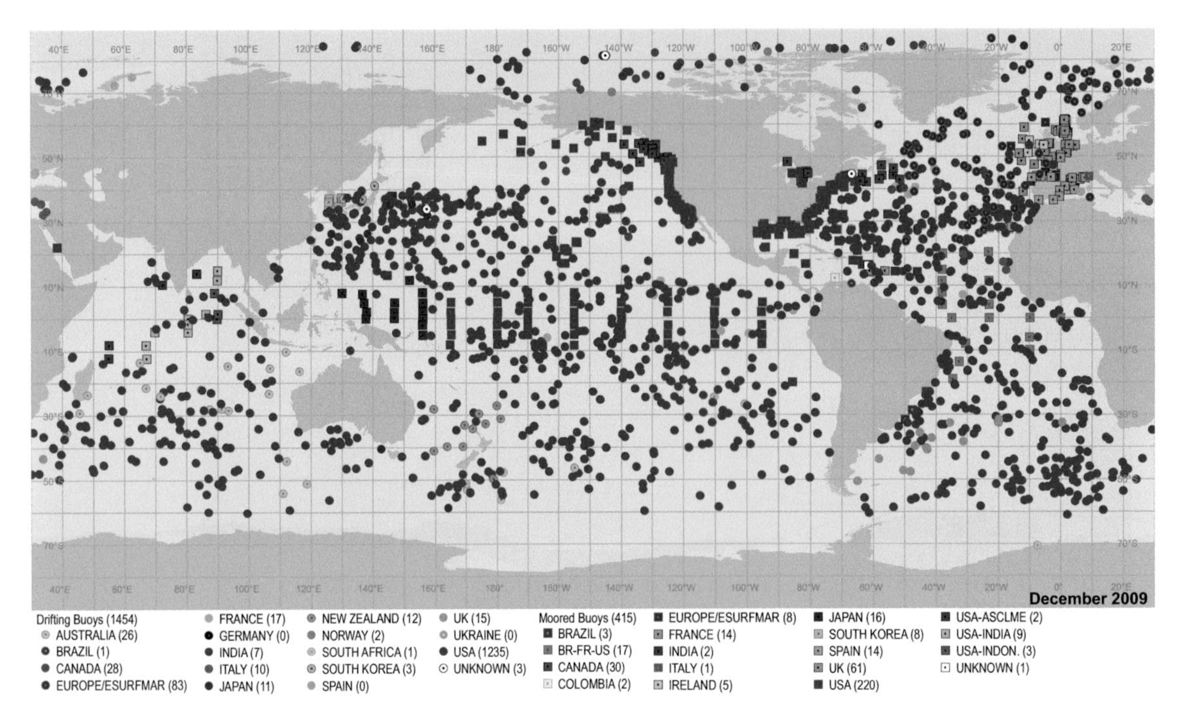

Figure 22.4 A map of the ocean data sources for December, 2009. Fifty years ago a similar map would have shown only a handful of points. © JCOMM.

The IOC is recognized as the competent UN body for ocean science in the eyes of UNCLOS and contributes to the elaboration and clarification of the application of UNCLOS, but arguably does not provide a stimulating or persuasive voice for enabling actions. UNCLOS, and other agreements such as the London Convention (1972) and the Convention on Biodiversity rightfully take precedence.

Despite these reservations, the IOC is clearly playing an increasingly important role in enabling and facilitating ocean science, ocean data collection and exchange, and ocean services, among other things. The response to the 2004 December tsunami shows that the IOC is capable of being proactive.

The 'Future of IOC' study examined institutional arrangements and concluded that '*The future of IOC should be based on the premise that the IOC remain, and be reinforced, within UNESCO. The IOC should look for an enhanced IOC role within UNESCO in terms of intersectoral cooperation, based on its technical expertise. Options for a more independent IOC outside of UNESCO were not broadly supported and recognized as being premature*'. This conclusion was reached principally because all the alternative arrangements looked out of reach, and not because

Figure 22.5 Deploying Argo floats for long-term ocean monitoring at depth. Argo is an automated observation system for the Earth's oceans that provides real-time data for use in climate, weather, oceanographic and fisheries research. More than 3000 floats report regularly by satellite link. © JCOMM.

Member States argued that UNESCO delivered significant irreplaceable additional value to the IOC.

Indeed it is quite difficult to find examples where UNESCO has proactively sought to use the IOC as a preferred mechanism for doing its work – the value of UNESCO seems to be more in the form of a steady hand from a UN family 'elder' ensuring the IOC can work effectively within the UN structure and maintain its 'form' and relevance. A future hope would be for UNESCO to become more active in using the capability of the IOC to deliver its programme.

Future expectations of science and technology

Trends in science

The scientific mandate and scope is another central issue. The IOC has recently worked hard to update its Medium-Term Strategy and to seek improved alignment with the objectives and goals of UNESCO. Interestingly, the conclusion was reached very quickly by the Future of IOC study that the Medium-Term Strategy was appropriate and had correctly captured the medium- and longer-term scientific challenges for the IOC.

Climate and climate change

Quite clearly, climate change is driving and shaping much of earth science at present. In recent decades, questions surrounding the nature and pace of climate change have dominated ocean science, including changes in surface and deep ocean temperatures, in regional and global sea level, in upper ocean and deep ocean circulation, in ocean acidification, in salinity and in El Niño; and the role of ocean modes like the North Atlantic Oscillation in climate change, among many other issues.

Though great progress has been achieved, the challenges of today remain as daunting to the current generation as did the issues being posed 20 years ago, when the first Intergovernmental Panel on Climate Change assessment was getting underway. Indeed the expectations, which include regional projections and climate change predictions, are placing enormous pressures on the present capability. During this time the role of the IOC has changed. In the past, the IOC was directly involved in coordinating and facilitating climate change ocean science. Today, the IOC's involvement is more oblique and has arguably less direct influence. On the other hand, the IOC now plays an increasingly important role in providing the underpinning infrastructure (e.g. observing and data networks) that support ocean science.

In the last decade, the focus for climate change has started to shift towards adaptation and mitigation. Climate change science remains important, but it is the implications for the coastal regions and living resources of the ocean that are starting to gain relatively more attention and prominence. The current IOC scientific strategy emphasizes this need, and the need to understand the implications of climate change for ocean hazards and extreme events. The majority of the IOC Member States have little or no capacity in the area of climate change mitigation, yet many of these are the most vulnerable to expected impacts. Over the next 30 years, during which climate change is inevitable, the IOC must accept a leadership role in facilitating and developing ocean science relevant to adaptation and assisting least developed countries, in particular small island developing states, to access and exploit scientific knowledge.

The marine environment and global change

Global change more generally, including that due to human activities, is another key factor that will shape the marine environment and intergovernmental science. Three-quarters of our planet's surface is ocean yet, compared to our exploitation and occupation of the land surface, we have done relatively little to use the marine space and have sought to exploit little of its total resources. Unfortunately, human activities have not left the oceans unscathed. We have indiscriminately used the oceans as a depository for our wastes. Urban, industrial and agricultural pollution flow down our rivers and streams and impact the shallow and productive coastal waters; we have altered the coastal habitats with developments that have taken a heavy toll globally on marine biodiversity; we have developed such efficient and effective technology for fishing that we seem to be rapidly denuding the oceans' wild fisheries; and, as noted above, through climate change and ocean acidification, we may be stressing the

Figure 22.6 An Argo float being deployed from a Japanese Research vessel. © Argo JAMSTEC.

marine and coastal environments in ways that will result in irreversible damage to marine ecosystems.

The immediate demand is for a system that will allow us to assess and develop an account of the state of the global marine environment; we need to know what is there before we can sensibly analyze change. Such a system will depend fundamentally on our ability to observe the ocean environment and measure change. The assessments will examine our understanding of the marine environment and analyze outlooks based on that understanding and the marine environmental information collected for the account. Over the next two decades there is perhaps no greater challenge for the IOC than to lead the development of such an assessment system within the UN, building on the success of the initial Assessment of Assessments phase.

Achieving future impact for the IOC

When one looks at these issues, and those surrounding marine ecosystem services and biodiversity, there would appear to be a major challenge for the ocean science community and the IOC. While we have had good success with facilitating and coordinating global scale programmes, including the current International Ocean Carbon Coordination Project and initiatives like GLOBEC, the science issues of the future are likely to be characterized by regional and local issues and demand a new type of approach. Member States continue to yearn for mechanisms that will enable them to access and take advantage of science and understanding at the regional level, which means regional coordination and development. The IOC like other UN bodies gets its primary efficiency and best net effect by optimizing at the global scale – modest relevance and impact for the majority rather than high relevance and impact for a few.

There is a need for a new science programme that is constituted from diverse regional and local elements, with the value-added being a common systematic approach to things like scientific error/uncertainty, data analysis and publication, and fundamental knowledge. The power to set scope and objectives would be vested locally and regionally to ensure the science was fit for purpose/need, and the programme would add value through the commonalities. The IOC, with its regional bodies and strong membership from developing countries, would seem ideally placed to provide such leadership and coordination. Moreover, if UNESCO was so inclined, there would seem to be opportunities to exploit the capability of IOC more broadly since actions at the regional and local level may be made more efficient by working alongside related UNESCO science initiatives. Regional subsidiary bodies such as WESTPAC provided opportunities to address this failing and in a number of cases they have been successful. Yet the progress and impact does not optimally exploit synergy with the global programmes, and suffers too often from lack of critical mass and scale.

Returning to the conclusions of the Future of IOC study, the above provides an alternative perspective, in essence arguing for a restructure of the regional model rather than perpetuating one that is destined to only partly work. There is also perhaps a suggestion in the discussion above that it will be events other than change in the direction of science that will influence the future. Emphases and relative priorities will almost certainly evolve, but there does not appear to be a branch of ocean science that is destined to grow from obscurity to dominance in the medium term. Indeed for climate change and global change more generally, the truth would appear to be that it is not the uncertainties and pace of progress of the science that is hindering the slow response of the world, but social and political will. Progress in ocean science is not the weak link, at least at the global level.

Trends in supporting infrastructure and technology

Optimism usually surrounds the development of technology and, if the evidence from the past 50 years can be used as a guide, technology will have a

Figure 22.7 Destruction in Aceh Indonesia in the wake of the catastrophic tsunami in December, 2004. Better preparedness would have saved many lives. An IOC tsunami warning system had long been established in the Pacific and international response to the Indian Ocean event has led to the establishment of similar systems in that region and other vulnerable areas around the globe. © A.S. Zain/Shutterstock.com.

profound impact on the future ocean governance. Technology will impact observing infrastructure, data collection and exchange, modelling and prediction systems, service delivery, and even the strategy for capacity development. It will provide new challenges and opportunities for coordination, as with ocean prediction; it will provide challenges in terms of transfer of technology from the developed states to developing countries; it will provide challenges in facilitation as in the adoption of uniform standards; and it will provide challenges in governance, as has been the case for the Argo floats network. The following represents only a few examples of this technology.

A maturing observing system

Previous chapters have told the story of the evolution of ocean observation into the Global Ocean Observing System (GOOS), and referenced the many challenges we face in sustaining that observing system. To some extent the IOC now sits at the crossroads, having taken the system from proof of concept through to its present state, but the community has yet to find a mechanism that will secure the system for the future. Progress with the coastal observing elements of GOOS has generally been disappointing, despite the preparation of excellent plans and the institution of novel coordination mechanisms, such as GOOS Regional Alliances.

Again, it begs for a different approach in the future. The IOC needs a new business model for the GOOS, one that removes excess intergovernmental process and procedure in favour of an approach that exploits the mechanisms that are working at the national and regional level. A solution may in part be through removing a layer of (intergovernmental) process and, perhaps even more radical, for the IOC to step up and take clear and explicit ownership and responsibility for the GOOS. The JCOMM would provide an assured path for coordination with the WMO and mechanisms other than joint sponsorship could be used to ensure effective coordination with the UN and international systems.

The dream of broad commitments to operational ocean observing networks is probably just wishful thinking. Rather, investment will be motivated by a range of needs, some related to research, others to high-priority routine needs of society, such as for tsunami warning. There is clear value-added in terms of intergovernmental coordination and standard setting and the IOC must focus its efforts on those areas that add maximum value. This can only happen if there is effective monitoring of the networks, of national and regional returns against plans, and of the value being harvested from the system.

As has been noted elsewhere, the lack of a Convention or similar instrument associated with the IOC and the lack of effective mechanisms at the national level are two key factors limiting the effectiveness of intergovernmental cooperation. The Commission needs to find a method, perhaps drawing in part on the reporting mechanisms used by the GCOS under the UNFCCC, which will allow regular monitoring and reporting of progress, and instil a greater sense of urgency and commitment to the plans of the IOC for the GOOS. GOOS networks, particularly coastal and non-physical elements, will not be sustained through the same mechanisms as operational meteorological systems; it will require a different approach.

The IOC must be prepared to play its part in ensuring the value of proven experimental and pilot systems is retained and secured for the future and accept that this will not happen by adopting the status quo. Through the JCOMM, it can claim a degree of success as many of the advances of the 1980s and 1990s observing networks were retained, but the IOC tends to place too much store on the intergovernmental power to make nations commit, and it falls to individuals and enterprising agencies to make things happen.

In terms of specific elements, securing the future of Argo is a challenge for the next decade, and should be given priority. As with satellites, it is often the attraction of technological advances that maintain investment interest, not the underlying socio-economic value. Paradoxically, it is often the implied long-term cost commitment that is a turn-off. The IOC can assist by removing barriers to efficient deployment and maintenance of float arrays and continually emphasizing and advocating the transformational nature of Argo observations for monitoring and predicting climate and climate change. Argo has been outstanding in broadening the depth of involvement of Member States in its operations and the IOC must

Figure 22.8 Skill transfer at work: getting some at sea experience on an IOC-sponsored training course. Scientists who train together can establish professional networks and personal friendships for life. © IOC.

ensure that Argo and all other elements of the observing system are available to all interested Member States and their value is made clear.

Information systems

At all governing body meetings of the IOC, there are strong expressions of support for IOC's data and information exchange work, and a sense of importance, purpose and belief in the growing strength of the systems. It is an area of strength of the IOC, albeit seemingly always shy of the real investment needed. Technology has transformed the way data are collected and exchanged and revolutionized information sharing. If the IOC is to meet its challenge of the global commons, information systems will be at the heart of the response, particularly in empowering the smaller and less developed nations to access information and knowledge.

The ocean community, along with some others, has struggled to take the leap to universal standards and protocols and one would hope this will be

Figure 22.9 The IOC now plays an increasingly important role in providing the underpinning infrastructure (e.g. observing and data networks) that supports ocean science. © JCOMM.

an early achievement of the next 50 years of the IOC. The community is in general agreement on where it needs to be, but it needs coordinated and concerted action that only an injection of additional resources can realize, which is hard to achieve in an intergovernmental world where downsizing and zero nominal growth is the norm. One would like to believe that the IOC in partnership with others of the international ocean science community could facilitate such a leap.

Some of the other big challenges lie with biological and biogeochemical data, and coastal science communities where there has yet to be uniform acceptance of the value of free and open exchange of data; or an understanding that the returns from freeing access far outweigh the costs and lost individual advantages of proprietary data holdings.

For marine environmental assessments and accounts, the power and utility of the underlying information systems will ultimately determine success. If barriers inhibit the quality and extent of the global information base, the assessment will lack credibility, the accounts will lack accuracy and outlooks may be impossible. The IOC must plan and work over the coming decades to ensure it enables the flow of information to all in a way that establishes the IOC and its data systems as world-renowned and respected sources of marine information.

Trends in future intergovernmental coordination and facilitation

The IOC has made significant decisions in areas such as data policy, transfer of marine technology and in consideration of the legal framework applicable to the collection of oceanographic data and, in particular, regarding the deployment of floats in the high seas within the framework of the Argo programme. Such matters require effective mechanisms for intergovernmental cooperation.

Figure 22.10 We have developed such efficient and effective technology for fishing that we seem to be rapidly denuding the oceans' wild fisheries. © Willem Tims/ Shutterstock.com.

The IOC and services

The IOC is one of the few parts of UNESCO that facilitates and provides services, and is associated with tangible and discernable benefits and outputs. In this respect it is more like the WMO and has led some, from time to time, to question whether UNESCO is the appropriate home for the IOC. The roots of IOC service provision can probably be traced to the creation of the Integrated Global Ocean Services System in the early 1980s, but this role has not flourished in perhaps the way it was envisaged at the time. The major reason is probably the general lack of national ocean service agencies among Member States. The evolution of the IGOSS into the JCOMM in 2001 provided a major impetus, but even today few readily identify the IOC with a major service function.

The 2004 Indian Ocean tsunami had a significant impact on the service side of the IOC so that today it plays a major role in the coordination of tsunami warning and mitigation. By most measures the IOC response over the last 5 years has been appropriate and effective. Sea level services provided through the GLOSS have continued to be strong, with services now coordinated through the JCOMM and its constituent programme areas. The GODAE has opened up the prospect of expanded ocean services to complement existing marine services, again coordinated through the JCOMM.

In terms of a more extensive range of services coordinated through the IOC the greatest demand lies in coastal areas and in responding to climate change. The unique capability that the IOC might bring to bear is its power to coordinate both globally and regionally, and its ability to connect to marine agencies, even if that connection has multiple and diffuse manifestations in many countries. To succeed would require lighter governance and greater efficiency than has hitherto been

Figure 22.11 The big challenge is to make capacity development tailored, sustainable and appropriate to the actual local/regional circumstances. © Eric Isselée/Shutterstock.com.

displayed by the IOC and its regional subsidiary bodies. It would require pilot demonstrations and a phased approach to develop confidence and surety. It would require a new modality for doing regional business, which respected national jurisdiction, standards, reliability and quality, and that used performance measures to proactively demonstrate the value-add of IOC coordinated services. It will require leadership from a few for the benefit of many. It should be a goal of the future IOC.

Facilitating the build of capability and the development of capacity

One of the pleasing advances in the IOC over recent times has been the development of effective strategy and implementation plans for capacity development. The strategy has been innovative and implementation has focused on enduring impact. The big challenge is to make capacity development tailored, sustainable and appropriate to the actual local/regional circumstances. In terms of the core work of the IOC,

Figure 22.12 An image of a local citizen taking up residence on a deep sea instrument. Image courtesy of NEPTUNE Canada.

capacity development must be fostered, even if that causes pressure on other areas of the IOC.

Capacity building in the IOC often struggles for critical mass and sustainability. It is a cross-cutting activity and as such is prone to overlaps and inefficiency and to being relegated in terms of priority. The Commission should set targets for its capacity development as part of its overarching performance metrics.

The leading edge of science and technological advances sometimes appears to be disappearing over the horizon as capacity development activities try to bring others up to date. One might ask 'Should the IOC assume responsibility for coordinating a particular area of science and/or technology if it cannot also deliver equitable and appropriate access to that science/technology for all Member States?' The suggestion here is that all IOC business plans should require a capacity development strategy and associated implementation element for all activities that come within its respective mandate.

The future environment

There are a few more general aspects to consider, including the political environment within which the IOC must operate. There are others who are in a better position to hypothesize on the future of the UN system, and UNESCO in particular. There are few if any among the developed nations seeking to expand the UN system, and many are content to restrict bodies like UNESCO to zero nominal growth. This is probably not unreasonable for core business elements where one could expect efficiencies and productivity gains to offset the diminished purchasing power of such scenarios. However, the lack of mechanism or the will to introduce new UN policy activities means there is a tendency towards business as usual and mediocrity, not enterprise and initiative. Within UNESCO, new activities must find offsetting savings, and for a small enterprise like the IOC this means it is almost impossible to maintain pace with changes and growth in its Member States.

Intergovernmental bodies like the IOC can generate extra-budgetary resources to offset these shortfalls and, indeed, the IOC has generally been growing in recent years through such resources and related in-kind and dedicated investments. The paradox is that this weakens the influence of the central intergovernmental process and vests influence in agencies and bodies beyond the UN system. The IOC at least has yet to conceive a budgeting, accounting and reporting system that properly takes account of all contributions to its work, and is far from a system of accountability and responsibility that would provide appropriate governance for all activities.

In this respect, the IOC would appear sometimes to be a vehicle of convenience or opportunity rather than a preferred mechanism for conducting ocean business. It lacks the efficiency and governance structures that are needed for effect and to attract activities to its operations. This can and should be changed, but it will require a willingness to exit from low-priority activities or those that really are not a good fit for the IOC.

The prospective areas of growth for the IOC and its services include marine environmental assessment, ocean science for adaptation, coordination of regional ocean science and related activities, and mitigating the impacts of natural hazards using warning systems. The IOC is not without competition in this marketplace, including from within the UN system. There needs to be respectful and strong coordination with the WMO, with acknowledgment of where each has its strengths and can lead, or where the role should be one of support. The partnership with the UNEP will be critical for the future and this needs concerted attention over the coming decade. The IOC does have a monopoly in a number of areas, but Member States will only exploit this route if they believe it will yield real added value.

In recent times there has been more than the usual interest in moving the IOC under different institutional arrangements, including a new hosting arrangement. Some nations have expressed interest in such a deal, particularly those with a strong marine interest. Such changes were discussed in the 'Future of IOC' debate but without resolution; one anticipates renewed interest over the coming decade.

Perhaps the biggest challenge for governments is to translate the ocean knowledge into information for use in the evolution of policy for the long term. It is in this area that governments must act cooperatively. Despite the transfer of jurisdiction over

Figure 22.13 The immediate demand is for a system that will allow us to assess and develop an account of the state of the global marine environment. An example of the insidious infiltration of our waste into the ocean is shown by this slide of microplastic debris collected at the surface in the North Pacific gyre. © James Leichter/Marine Photobank.

large areas of the ocean to nations, the international commons of the remainder is still vast. It is impossible to manage the global ocean on a piecemeal basis or to consider that actions undertaken in one part would not have an impact elsewhere. Governments have gradually accepted the scientific understanding that human intervention is changing our atmosphere and threatening the future. The same acceptance and consideration must be given to the oceans and will be even more difficult, but equally critical for our survival. The policy relevance of IOC activities needs to be strengthened but this is difficult with its present location in UNESCO; it is simply out of sight for most national policy development.

Synopsis

The future of IOC will not be shaped by a vast winning of influence and resources in a 'UN Lottery', but by Member States working together towards a shared aim. The IOC enjoys respect within UNESCO and generally through the UN system, but lacks recognition and impact at the national level. Few of the next generation of ocean scientists have an appreciation or respect for intergovernmental processes, and perhaps that is as it should be for young scientists. But for ocean science, the IOC is a necessary element of the fabric of ocean endeavour; but it should mostly be content with delivering value in its area of competency and not seeking to lead ocean science more generally.

This paper suggests that the future of the IOC is mostly within the control of its Member States, not external forces and drivers. Science and technology will evolve but probably not in a way that will fundamentally change the way the IOC operates. Rather they will provide opportunities and challenges for evolving the products and services coordinated through the IOC and its partners.

The IOC can do much to prepare itself for the next decades, through introduction of innovative coordination mechanisms (light on governance, heavy on regional relevance and impact) and improved governance arrangements. The latter is critical if the IOC is to become the preferred mechanism for coordination and facilitation of ocean activities, regionally and globally.

The UN still lacks effective UN agency coordination mechanisms for ocean matters and the IOC presently lacks the influence and capacity to lead such coordination. The IOC is not optimally located within the UN system but its location as an element of the Natural Sciences Programme of UNESCO is not likely to change in the near term.

Ocean policy makers and scientists should look to the future with confidence, but also be prepared to support change in the way they cooperate, and specifically how they use the IOC. The oceans matter and the Intergovernmental Oceanographic Commission can and should be central to the global effort to manage and sustain the ocean for the benefit of future generations.

23 Afterword

This book was conceived as part of the 50th Anniversary celebrations of the Intergovernmental Oceanographic Commission of UNESCO. The Member States of the IOC planned the celebration with three objectives: to record the past, to tell the world about present activities and to confirm commitments by the Member States to the future of the oceans. As our role as editors developed, so too did the direction and emphasis of the book. The scope has extended to the widest issues of oceans and the role of science in helping governments to manage them wisely, consistent with addressing as broad an audience as possible about the needs and issues that exist and must be tackled in the future. In these final pages, with the strong encouragement of our publishers, we attempt to draw together several of the themes which have emerged in the individual author contributions. Perhaps the most elusive objectives are those that connect with the future: how will the oceans change and what is the role of science?

Some predictions are more probable than others. Continuing rises of sea level are very probable, as are increased acidification and warmer temperatures. Contamination by chemicals and the discarding of other waste materials will continue in the short term, and longer if strict controls are not imposed. Oceanography will help deliver long-range forecasts of weather and climate, which will enhance the global economy. But, as with all science, the directions and discoveries for future ocean science are intrinsically unpredictable: a famous oceanographer once said that if he knew what he would be doing in 5 years time, he would be doing it now. Given that major marine experiments need several years to plan, that may be slightly misleading, but 10 or 20 years ahead is unknown scientific territory.

To put the correct perspective on any predictions of what the future holds for the oceans, it is as well to reflect on the rate of change of our technology over the past century. Imagine a chapter being written 100 years ago about the future of air travel for example, or even 50 years ago on satellite and space technology. Are such comparisons anomalous? The oceans have been part of our history and culture throughout the existence of our species, and yet we are a terrestrial society. We have carved our civilizations on land, with our cities, roads and agriculture so that terra firma no longer resembles its natural state. Yet three-quarters of our planet's surface is

Figure 23.1 Will ocean farming become a reality? Inside the cage on a tuna farm. © Marco Carè/Marine Photobank.

ocean and over this same 100-year period we have done relatively little to constructively manage and use marine space and its resources. Unfortunately, this does not mean that our land-based practices have left the oceans unscathed. Without considering the consequences, we have indiscriminately used the oceans as a depository for our wastes. Urban, industrial and agricultural pollution flow down our rivers and streams and impact the shallow and productive coastal waters. We have altered the coastal habitats with developments that have taken a heavy global toll on marine biodiversity. Probably most seriously, industrial practices have changed the composition of our atmosphere and have led to changes in radiation, and increased warming and acidification of surface waters and consequential threats to marine ecology.

So, in this context, is the future for the oceans one of doom and gloom? Perhaps not, but one thing is absolutely certain. As a society we must collectively and unambiguously acknowledge the importance of the oceans to our existence on the planet. The oceans are the lungs of the planet: they influence our weather and climate, and are the major part of the hydrological cycle on which we depend. We must be aware of the changes we bring to the ocean and the consequences of our actions. For this we require information and the knowledge to use that information wisely. To return to the analogy of our terrestrial evolution, as our numbers increased we had to turn from supporting ourselves with hunting practices to agriculture: over the years we gradually adapted our landscapes and practices to more intensive production. At present in the ocean we are still hunters, but with our new capacity in technology and sparse international controls we seem to be rapidly denuding the wild fisheries. Is it too much of a stretch of the imagination to believe that it will someday be possible to farm the seas and its renewable resources and increase its productivity manifold? Would a hunter many centuries ago been able to predict the development of the farms and agriculture that we have today? The creation of fish habitats is not a new concept. Ships have been sunk to provide artificial reefs, but as yet no-one has considered

Figure 23.2 A polar bear and her cubs, a species threatened by global warming, take exception to an instrumented buoy imbedded in the Arctic ice to measure changes. Image courtesy of D.G. Barton, © 1992.

shallow floating fish habitats on the size of terrestrial farms, seeded with weeds and sea grasses and harvested sustainably for the fish stocks that will naturally gather or be introduced there.

What of ocean space? As sea level rises and threatens our coastal and island habitats, it would seem natural to consider how modern society will cope. One obvious solution would be to create more space with enclosures, artificial islands and even mobile habitats that could combine renewable energy sources, wind, wave and even hot vents, with self-supporting food production. In many countries, the sea is used as a source of fresh water. Presently, this production of freshwater from the sea has a high energy cost, but already there are engineering and demonstration projects to distil freshwater using ocean energy sources that will reduce costs and provide a greener solution.

Marine transportation has been a critical part of society for many centuries. Today 80% of our trade goes by sea. There have been major technological changes to our fleets; ships the size of small villages carry our goods and people from place to place around the globe. There may yet be surprises in transportation evolution. Certainly the environmental footprint of the present marine shipping and its shore-based facilities needs to be reduced, but whether there will be new forms of transport, perhaps even a return to sailing ships, that will revolutionize maritime travel remains to be seen.

Another imponderable is the vulnerability of the deep ocean itself. Very recently researchers have found life *within* the sea floor. These life forms, such as bacteria, can withstand the enormous pressures and temperature ranges found at ocean depths. It is expected that half the world's biodiversity exists in the ocean depths and is yet unclassified.

With governments coming to terms with the need for renewable energy, the oceans may contribute to the solution through wave and tidal power, surface winds and

Figure 23.3 The biodiversity of the deep ocean is only now being explored. A dense bed of hydrothermal mussels is seen near a seafloor hot spring. Other vent animals living among the mussels include shrimp, limpets and Galatheid crabs. © NOAA.

subsurface currents, thermal and salinity gradients, and even geothermal energy. The cost of such enterprises may be high, but when the survival of the planet is at stake the economics assume a less important role.

The task of marine science is to provide governments and managers of coastal communities and marine activities with the information they need to give an economical, safe and sustainable service to society.

- Economical, because priorities for ocean services have to compete with other societal benefits and they must be shown to be worthwhile.
- Safe, because the natural ecology and our own life and property must be protected.
- Sustainable, because protecting the health of the planet and the health of the ocean are one and the same goal.

A proper scientific understanding is the basis for wise management and sustainable use. Monitoring changes is essential. We have made great strides over the past two

Figure 23.4 Marine life is fascinating for all ages. © BlueOrange Studio/Shutterstock.com.

decades in the quantity and quality of ocean observations. The use of satellites has given ocean scientists much greater spatial coverage, sea level data and surface characteristics. Automated surface and subsurface floats, communicating by satellite, have transformed our knowledge of the water column and the ocean processes taking place. We now have the capability to implant fish and marine mammals with transmitters so that their migration routes and habitats can be studied. The deeper ocean is still relatively unexplored, but underwater observatories that will bring video links and sea floor information directly to shore-based laboratories through fibre optic cables are one recent innovation that will help.

There will be continual changes to technology that will increase the quality and quantity of observations and with it opportunities for new and more accurate and relevant predictions. More importantly there will be great changes in the way we use the ocean and its resources and the corresponding need for increased knowledge and information services.

The world, and individual nations, will need more ocean scientists to receive and understand this new flood of data and turn it into information that can be used by

Figure 23.5 A child in Mauritania goes home with fish. © Demi-UNEP/Still Pictures.

decision-makers and managers. Governments will need to support education and research facilities to make this possible. In addition the infrastructure that links the science to the information services needs to be in place. In this regard it is useful to compare future ocean services with the climate and weather services that exist today for meteorology but are only in a fledgling state for the oceans. Marine surface operational information for shipping on tides, winds, waves, currents and even ocean temperatures are sophisticated and well used. Fishing management information, or at least the use of such information to sustain fisheries, has not been entirely successful and more needs to be done. Coastal hazard warnings to alert and mitigate the impact of tsunami, storm surges and wave action are becoming more commonplace. Coastal area management covers a wide variety of conflicting marine activities in the near shore areas and information services for successful regulation are not yet sufficient in most areas.

The human race is no different from any other species in being willing to adapt their environment to better serve their needs. But we are unique. First we have the technology to make changes large enough to affect the global environment and second, perhaps even more seriously, we rarely act together on a global scale. We humans are too often driven by motives of short-term individual or corporate gain. Even national interests can transcend the more logical collective approach to long-term stability and survival. The two characteristics taken together make for a dangerous situation that needs to be recognized and addressed by our respective leaders.

The biggest challenge is to translate the ocean knowledge into information for governments to use in the evolution of policy directions for the long term. It is in this area that governments must act cooperatively. Despite the transfer of jurisdiction over large areas of the ocean to nations, the international commons of the remainder is still vast. It is impossible to manage the global ocean on a piecemeal basis or to consider that actions undertaken in one part would not have an impact elsewhere. Governments have gradually accepted the scientific understanding that human

Figure 23.6 A bridge made of fishing gear recovered from a protected 'no fishing area' in Costa Rica. © Todd Steiner, Sea Turtle Restoration Project/Marine Photobank.

intervention is changing our atmosphere and threatening the future. The same acceptance and consideration must be given to the oceans and may be even more difficult, but is equally critical for our survival.

A world renowned author of science fiction, Michael Crichton, in a preface to his book entitled 'Prey' argues for a precautionary approach:

The fact that the biosphere responds unpredictably to our actions is not an argument for inaction. It is however a powerful argument for caution, and for adopting a tentative attitude toward all we believe and all we do.

Intergovernmental cooperation is essential, but intergovernmental meetings alone will neither develop needed policies nor provide the catalyst for change. One of the most important lessons from the chapters in this book, and from our own experiences is that while meetings are the vehicle, it is the energy, creativity and dedication of the individual that provides the driving force. Attendance at a meeting leads to people committing and vigorously working for progress, and implementation back home. The commitments generated, and the feeling of ownership that true cooperation engenders, are essential, as shown in several chapters in this book.

Perhaps the greatest uncertainty in predicting the 50-year future for the oceans is the role of governments in actively and collectively managing our common resource. The United Nations system has tried, with limited success. As an institution founded in the middle of the last century, at the high-water mark of the absolute sovereignty of the Nation State, however small, it works painfully slowly, and in always seeking to work by consensus is too vulnerable to the perceived short-term interests of individual States. UNCLOS is still an imperfect construction that needs refining based on experience. Several key States have not yet ratified, and there is no effective way of monitoring or policing its implementation. Wars have been fought to protect and exploit marine resources, and they will again be fought in the future.

Figure 23.7 We want the next generation to be happy with our legacy of a healthy ocean and its bountiful resources. © Rebecca Weeks.

The way ahead is uncertain and, in terms of the planetary environment, could be likened to a minefield. An ignorant person may march straight ahead and seek to avoid danger by chance alone. A wise person acting with caution is much more likely to survive. The size of our footprint can be reduced by environmental actions, by tackling poverty, ignorance and population issues and working through governmental cooperation. Wise use of technology can help and conversely mismanagement of technology could hinder. Increased knowledge can assist in avoiding larger problems, in mitigating unavoidable natural dangers and sometimes, in recovering from our mistakes.

Here, then, is a final synthesis of the several strong messages that emerge from the authors of this book.

First, the oceans are an essential part of life on the planet. The physical, chemical and biological processes of the ocean supply water to the hydrological cycle, absorb carbon dioxide from the atmosphere, cleanse the air we breathe and distribute the heat around the globe. For the human race, the oceans provide resources, trade routes and a repository for our wastes. There is evidence that the ability of the ocean to absorb the wastes of our society is being exceeded in many coastal areas; there is also evidence that the surface waters of the ocean are becoming more acidic, due to the increase in carbon dioxide in the atmosphere. We should be concerned, because we neither fully understand these changes nor how we should act to protect ourselves from potential dangers.

Second, science encompasses observations, knowledge and information. The relative amount of resources spent on ocean science is very small, partly because of past difficulties and the expense of collecting ocean data, partly because we are a terrestrial species and partly because the oceans themselves have been regarded as vast and unchangeable. More knowledge of the ocean is needed to understand our environment and its potential changes and that knowledge needs to be transformed into information for managers and decision-makers.

Third, as the issue is a global one, governments must act together. Cooperation amongst governments leads to more efficient and effective programmes. Logically global issues should take precedence over national and regional priorities, although that logical transition is hard to achieve under present intergovernmental mechanisms.

The oceans have been part of the geological evolution of planet earth for hundreds of millions of years. Against a geological time scale the past 50 years is infinitesimally small; and measured in terms of chronology, so too will be the next 50 years. However, one prediction is certain: for a healthy and viable future ocean, the next 50 years will be the most critical of all time.

Index

Printed in the United States
by Baker & Taylor Publisher Services